Lumbermen on the Chippewa

by

Malcolm Rosholt

ROSHOLT HOUSE
Box 104
Rosholt, Wis. 54473

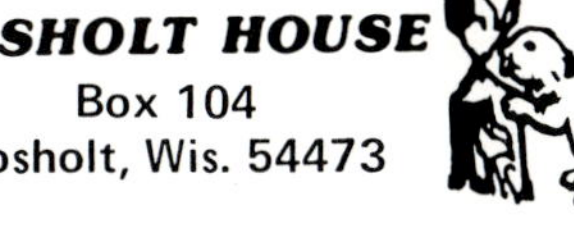

Foreword

In the preparation of this book I have been generously aided by Alberta Rommelmeyer, John A. Stoneberg, Dale A. Peterson and Dr. R.C. Brown, all of Eau Claire, Wisconsin; Joe Joas of Chippewa Falls; Eugene and Dolores Harm and Joyce Gannon of rural Cadott; Howard Peddle of Iron River; Ed Brick of the Department of Natural Resources, Madison; and by Michael F. Slasinski of Saginaw, Michigan.

For editorial assistance I am indebted to Beulah Larson of Stevens Point and to my daughter, Mei-fei Elrick of the University of Guelph, Ontario, Canada.

I wish to thank Mrs. Leslie Jones of Holcombe for bringing to my attention the daily journal kept at the Little Falls dam, and for introducing me to Zac Jardine of Verona who, as a youth, watched the last pine log sluiced through the dam in 1911. His sketches on how the "splash planks" were used in the dam have been enlarged by students of cartography at the University of Wisconsin-Stevens Point under the direction of Ray Specht. The students also did several maps in the book for which I am grateful. Finally I want to express my appreciation to Randal E. Rohe of Menasha for the diagrams of the up-and-down sawmill, the Tainter gate, and the bear trap sluice gate.

I recall with nostalgia the many people who have donated pictures to me over the years, people I talked to about working in the woods and on the rivers and in the mills. One of the first who come to mind is the late Hans Lee of Rosholt who drove logs on the Little Wolf River from the "Upper Dam" in Marathon County to Boom Bay on Lake Poygan. My old friend, Nils Quisla, who lived east of Rosholt, I visited many times on the family farm and listened as he talked about his days in the sawmills and in the woods. He vividly remembered the fear among the superstitious lumberjacks on the night of Halley's Comet in 1910.

Chester Colpitts of Radisson worked on the landing one winter on the Thornapple for Chippewa Lumber & Boom Company and recalled the struggle between John F. Dietz and the boom company at Cameron Dam. And Walter Bunk of Eau Claire told me about the winter he worked on the Thornapple for Moses & Gaynor moving logs by steam hauler to Babbs Landing on the Flambeau.

Henry Arnold of Green Bay remembered how the icing tankers were used in Upper Michigan and, by inference, in Wisconsin. Peter Brzezinski of Portage County, a head sawyer for Mohr-Stotzer Lumber Company at Holt—now a ghost town in Marathon County—remembered John Griffith, the mill foreman, whom he liked because even though he had a "bad temper" he "never held any grudges and he was an honest man."

Bob Mears of Mercer spent much of his life in the woods and what he didn't know about logging wasn't worth a whistle. John Mortenson of Aniwa was one of the last jobbers for Hines Lumber Company on the Lac Court Oreilles Indian Reservation in the 1930s.

William Hoyer of Cudahy, born in Chippewa Falls, I quote at length in the *Wisconsin Logging Book* (1980) and briefly in this volume.

The present volume takes advantage of the wealth of pictorial material found in the Chippewa Valley, most of it previously unpublished. In several instances I was forced to borrow pictures from the *Wisconsin Logging Book* because of their uniqueness, for example, the photograph of the fin sheer boom taken at Alma which appears to be the only one of its kind in Wisconsin.

Finally, I wish to thank Tony Wise for the privilege of photographing the picture of the log driver which appears on the cover of the dust jacket. It is a detail taken from a large mural done by Lyle Roger Nelson for one of the restaurants at Telemark Lodge, Cable, WI.

The photograph on the back of the jacket I took at Read's Landing, MN May 28, 1982. The view looks east across the Mississippi River to the mouth of the great Chippewa.

Malcolm Rosholt
Rosholt, WI 54473

Library of Congress Catalog Card No.: 82-060957

ISBN: 0-910417-00-8

Printed in the United States of America by
Palmer Publications Inc., Amherst, WI 54406

Contents

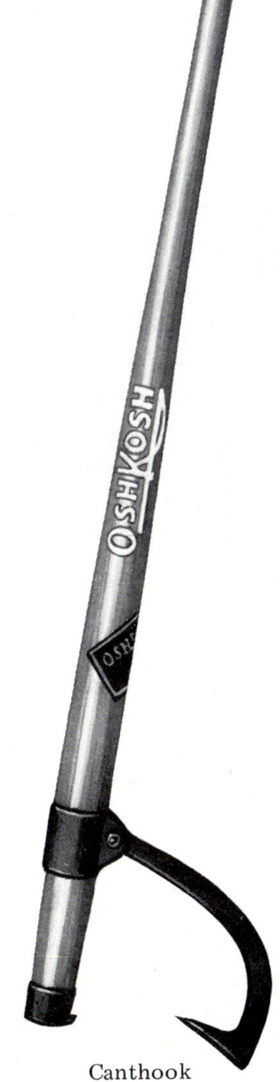

Canthook

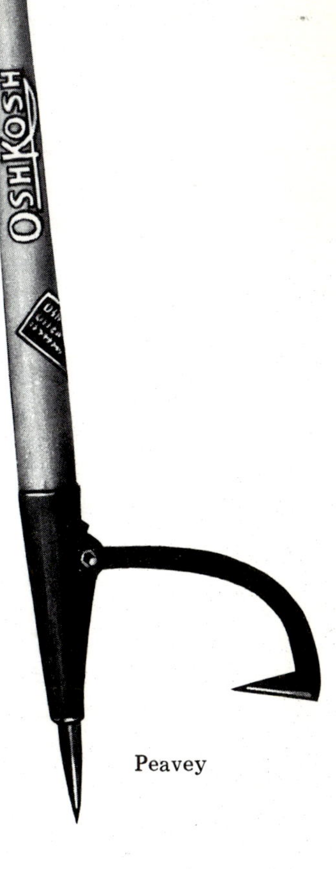

Peavey

The River

Chippewa River below Round Hill

It seems beside the point to say that the Chippewa Valley once contained billions of board feet of pine timber. (A board foot is a square board, one foot long, one foot wide and one inch or less in thickness.) It seems more to the point to say that the Chippewa Valley once held the greatest number of white pine in any area of equal size in the United States.

Despite these rather startling statistics, the total production of lumber in the Chippewa Valley for the year 1881 amounted to only 1.66 percent of the total estimated production of lumber in the United States, according to the U.S. Department of Agriculture.

In most forest land of Wisconsin, there were "openings land," that is, forties of land where there were openings in the forest created either by nature or by forest fires. And there were forties covered mostly with brush and occasionally a twisted oak, or willow leaning over the banks of a stream glutted with old windfalls and brush. But in the Chippewa Valley the forest took up five-sixth of the total land area, interspersed by rivers, lakes, swamps, and the Blue Hills.

It was to this heavily forested region that the first lumbermen came to live in the 1820s and 1830s, to cut down trees, buck them up, skid them to a river and float them down to a sawmill and saw them into square timbers, planks, boards, studdings, shingles, laths and pickets.

Since the timber in the woods could be purchased cheaply, or, as in many cases, stolen from the public domain, it is often thought that most lumbermen became wealthy. Look at the beautiful homes they built in Neillsville, Chippewa Falls, and in Eau Claire. On the contrary, the history of the industry in Wisconsin reveals that lumbering was a risky business in which more men failed than succeeded either as sawmill operators, or as loggers. There seemed to be no end to the natural disasters that lumbermen could fall heir to either by fire in the timber, fire in the mill, or floods, or open winters with not enough snow for good sleighing. Scarcely a year passed without some log boom breaking open sending thousands of logs downstream in a mad stampede. Many drifted ashore and had to be rolled back into the river with horses or oxen. There was a constant struggle to avoid log jams because they cost mill owner delay and added expense.

The early lumbermen also contended with the financial panics of 1857 and 1873, and with "Cleveland times" in the 1890s, and the panic of 1907 when nearly all the mills and camps were shut down.

Any trade, however, attracts entrepeneurs, men of little experience but lots of courage. Yet, without proper management techniques and

knowledge of the market for the products they manufactured, these plungers often lost their investment at the slightest setback. It was that way in the lumber business; too many men, attracted by what appeared to be easy profits, learned to their sorrow that there were no easy profits and that, without sufficient capital to tide them over a fire or flood, they were no match for the great white pines glaring at them from across the river.

Much of the lumberjack nomenclature heard in the Lake States of Michigan, Wisconsin and Minnesota was brought here by the lumbermen from Maine and lower Canada who had run out of pine and moved west. Among the interesting expressions they brought with them was "good chance" and "bad chance." A good chance was a tract of timber which might prove profitable; a "bad chance" was one where the risk factor was out of proportion to possible gains. These two expressions seem to epitomize the ups and downs of the lumber industry in the 19th Century. Interestingly, the term "chance" is still used in federal land transactions.

The hand of man and the whims of nature have changed the Chippewa River in the past century, yet, whatever the changes, it continues to form the watershed for nearly one-sixth of the state of Wisconsin and drains the counties of Pepin, Eau Claire, Dunn, Chippewa and Rusk, and parts of Buffalo, Pierce, St. Croix, Clark, Taylor, Barron, Washburn, Sawyer, Bayfield, Ashland, Vilas, Oneida and Price. The water from all these counties join the Chippewa at one point or the other and flows southwest into the Mississippi River below Lake Pepin and thence to the Gulf of Mexico.

The river below the city of Chippewa Falls slows down and can easily be used for lumber rafting, or floating logs. The exciting part of the river once lay between the Falls and mouth of the Flambeau, sixty-eight kilometers (42 miles) or so as the current flowed, and this was referred to in the lexicon of the old river drivers as the "wild Chippewa," for here the river tumbled and pitched down a slope which fell more than seventy meters, through falls once named Little, Brunet, Jim's and through rapids named Paint Creek and Eagle. But all this has changed and that is why it is described in the past tense. The river is almost one long lake today, created by a series of dams extending all the way north from Chippewa Falls to the mouth of the Flambeau.

Jim's Falls, now called Jim Falls, was named after James Ermatinger, an immigrant from Canada whose family settled at the falls in the mid-1840s. Brunet Falls, named for Jean Brunet, was located a short distance upstream and in the logging era a perennial source of log jams. The river at one time was forced into a narrow channel through the rocks and unless the logs were given a last minute shove with a pike pole to keep them running straight, they often turned crossways and started a jam.

Important to the lumbermen above Big Bend in the Chippewa River were several rapids which the drivers had to contend with such as "Buffington's," "Devil's Creek," near the creek of the same name; "Grand Rapids" about ten kilometers above the mouth of the Thornapple; "Nine Mile" or "Long Rapids" below the mouth of the Brunet River, and "John Hermon Rapids" just above the junction of the Couderay and Chippewa Rivers in modern Sawyer County.

The Flambeau River is the largest tributary of the Chippewa, and about eighty kilometers north of the mouth it divides into two forks, the south, often referred to in early days as the "Dore Flambeau" (a name retained in part in Lac Sault Dore), and the north fork.

The most common sources of log jams on the main stem of the Flambeau, that is, below the forks, were at Cedar Rapids, and at Big Falls, north of Ladysmith. Here the river pitched at least twelve meters in three successive rapids. At Muskelonge Falls (today Park Falls), the north fork fell about ten meters in three kilometers. Other rapids on the main stem of the Flambeau were known as "Shaw's Rapids," after Daniel Shaw, "Vinette Rapids," after Bruno Vinette, an early trader and lumberjack, and "Josie Island Rapids," probably named for Malcolm Josie and after whom Josie Creek on the Flambeau is also named. None of these names appear on modern maps of the Flambeau River State Forest.

West of the city of Phillips runs Elk River which joins the south fork of the Flambeau twenty kilometers above the forks.

Two other important log driving streams in the late 19th Century were the Brunet and Thornapple which flow southwesterly out of Sawyer County, the former joining the Chippewa near Exeland, and the latter joining the Chippewa west of Ladysmith. Neither one of these rivers seem significant today, but when their water level was raised by dams, and when the windfalls and alder brush had been cleared away, they were both important log driving streams. Early dams on the Thornapple were Shaw and Hoyt and, in 1877, the Leavitt dam, after Thomas F. Leavitt who built it. It was later known as Cameron dam after Hugh Cameron, site of the much publicized struggle between John F. Dietz and the lumber companies.

Another important log driving stream, an affluent of the Chippewa, was the Jump River which flows southwesterly and joins the Chippewa below the Big Bend. Today, the backwater of the Holcombe flowage covers the old mouth of the Jump. Log jams were frequent on the Jump, particularly around Devil's Nest which was located about ten kilometers northeast from the mouth in Section 1 of Holcombe township.

The last important tributary of the Chippewa is the Yellow River where there once were rapids named Cadott, Colburn, Slap Jack, Devil's Rock, and Coldwater Rock.

From north to south, the best known smaller streams feeding the Chippewa on the left (east) bank are Nail Creek, Deer Tail Creek, Fisher River, and Paint Creek; on the right or west bank of the river are Soft Maple, Potato, Devil, Bob's, O'Neil and Duncan creeks.

By 1882, it was estimated that there were no less than thirty dams on the Chippewa and its tributaries above Holcombe and there were more built later. In addition, there were dams below Little Falls on both the Chippewa and its tributaries such as the Yellow and Fisher rivers and Paint and Duncan creeks. Some of these were used to sluice logs, and others were "splash" dams, that is, when more water was needed on the Chippewa to create a "head", or flood, the splash dams on the smaller streams were opened.

The Camp

Scene at one of Eau Claire Lumber Co.'s Logging Camps, North Fork Eau Claire River, Winter of 1879--80.

THOS. DAILEY, - Foreman

Caption above appears on original picture. It may be the oldest, authentically dated picture of logging camp in Chippewa Valley. The photographers were Tebo & Hanson of Eau Claire. Bunk camp is of logs with pantile roofing, probably basswood hollowed out and laid like Chinese tile. At the left is probably stable for oxen and horses. Man at right stands by shaving bench; two men in center stand behind a go-devil; cook and cook-ee stand at door and youth at left is probably chore-boy. Empty barrels may have contained salt pork consumed during winter.

In the last decades of the 19th Century and first three decades of the 20th Century, many men looking for work in Wisconsin went "into the woods," or, they said they were "going up north." They were once called "shanty-boys," a term replaced in the 1890s by "lumberjacks." These were the men who chopped the trees, sawed them into logs, and skidded them behind horses or oxen to a skidway where they could be loaded on sleighs and hauled to a landing on a river, or to a railway siding to be shipped out on flat cars.

Cutting down trees was big business. There were millions of trees to be cut and there was an almost unlimited demand for forest products including everything from square timbers, ship masts, lumber, studdings, shingles, pickets, barrel staves, tools and implements. Even some roads were made of planking.

Although the forest was filled with millions of trees there was one which the lumbermen preferred above all others, the white pine; it was light and yet tough. In northern Wisconsin, roughly north of Highway 54 today, nature planted these pine trees in abundance. There are pictures from this early period showing pines standing one up against the other, so thick that no daylight can penetrate them. These pictures are accurate as far as they go, but not all land was equally suited to the heavy growth of pine. There were swamps, wetlands, hills, pot holes, and there was clay soil and light soil where other trees such a hemlock, yellow birch, tamarack and cedar grew better if not so abundantly.

Many photographs taken in the logging era show pine logs of extraordinary size. These logs were selected especially for the benefit of the photographer, an early form of boosterism, but they did not represent the average size. The average can be seen in photographs of logs decked on rollways where they are probably about forty centimeters (15¾ inches) in diameter at the top end.

The procedure following in cutting the trees was much the same in Maine as in Wisconsin and Michigan, since it was the New Englanders who brought their experience to the Lake States and taught the locals how it was done. In the fall, a small crew went ahead of the main crew to build a camp where the lumber company had pine to cut. The men carried a tent and were accompanied by a cook. They brought axes, crosscut saws, grub hooks, shovels and all manner of carpenter's tools. Their task was to cut a road into the future camp, cutting down the brush and trees and smoothing out a surface. The road had to be level, avoiding swamps, creeks and hills as much as possible. When possible, a camp site was picked near a spring or river where the cooks would not have to walk too far for their water supply.

The crew cut pine trees to build a camp which, in the 1860s and 1870s, was called a "bunk camp," presumably because it was a log building with bunks, not beds. Later, when camps were enlarged to include individual buildings for the men to sleep in and to eat in, the buildings were referred to as "shanties," i.e. the "sleeping shanty" and the "cook shanty."

The big camps always had to have a "hovel" or barn for the horses and cattle. The latter name referred to oxen, not milk cows. Most camps kept a milk cow, not for milk to drink, but for cooking purposes. Nor was cream ever served with tea or coffee; at best it was skim milk.

An early photograph taken of a camp for the Eau Claire Lumber Company in 1879-80 shows a low, almost flat-topped building of logs which combined sleeping and kitchen quarters. On the roof, instead of shake shingles, there are long pieces of hollowed basswood which have been laid down alternately, one up, one down, like tile on a Chinese roof, This was called pantile roofing, but no matter what kind of roofing, there was seldom any insulation in it and much of the heat from the stoves escaped through it.

Roofing for the very early log buildings varied. Bruno Vinette, describing camp life in the 1850s on the Chippewa River, said the shanty he lived in was covered with shakes, that is, shingles of pine more than a meter long which were laid on poles running lengthwise to the roof. The shakes were held in place by other poles on top, and the whole covered with clay. When the clay froze, the roof was covered with spruce or hemlock boughs to keep the clay from thawing in mild weather.

Early bunk camps used an open fire for cooking. French-Canadians called it a *cambuse*. When Bruno Vinette first came to work in a camp on Chippewa River in 1855 he said the cooking was done on an open fire. Smoke escaped through light in roof, and this was usually the only daylight in building. Documentary evidence proves this type of open fire for cooking was still used in a camp on Wisconsin River as late as winter of 1870-71, but it was one of last. Most camps already were using stoves in kitchen and crude heaters in sleeping shanty. Crane at corner seen in picture above held big pot. Fire burns in a sand pit, and at night the cook often buried a pot of beans to simmer. Long shovel held by cook was used to turn logs and to remove ashes. Socks and towels hang on string at upper right. Picture was taken on Ottowa River in Canada in 1900.

Early bunk camps were not much larger than twelve by twenty feet and constructed of logs, usually pine. Inside the shanty, the sleeping bunks were built with the lower end, or foot, facing inward toward the fire or cook stove in the center of the building. The bunks were made of boards covered with hemlock boughs or straw, but in the decades after 1900, mattresses were introduced, and finally by the 1930s, individual spring beds were not uncommon. (The author slept in a single board bunk, with mattress, on board a house boat for the Eagle River Lumber Company in 1926.) The new style of sleeping shanties after 1880 had bunks running parallel to the wall, usually two tiers high.

The sleeping shanty of the 1880s and later, had a heater, or oil barrel made into a heater, in the center of the building, usually too hot for the men near the stove and too cold for the men at the ends of the building.

But logging camps before the Civil War had no stoves. Taking their cue from the Indians, the loggers built a fire place in the middle of the room with a light on the roof for the smoke to escape. The French word for this kind of fire place was called *cambuse*, a word which Canadians have corrupted to "camboose" and which William W. Bartlett, ghost writing for Bruno Vinette, misunderstood as "caboose."

The *cambuse* was built with four posts sunk into the ground separated at the corners by about two meters and braced across the top with poles or planks. On one of the posts was a crane from which an iron kettle was suspended over the fire. The fire was built on a mound of sand about half a meter thick and this helped keep the heat in and also provided a place to dig a hole for a pot of beans to simmer in over night. The cook stands by, holding a long shovel to turn the ashes or to turn the logs in the fire. Most of the smoke escaped through the vent in the roof, but often when the wind was from the east, the smoke beat back and filled the camp with a blue haze. Usually, there were no windows, only the light from the vent, and even after larger sleeping shanties were introduced, only one or two windows were framed lest any heat be lost.

Although the *cambuse* was being used in the United States down to the early 1870s, most camps by that time had converted to stoves for the cook shanty and heaters for the sleeping shanty. In Canada the *cambuse* was used, at least on the Ottowa River, as late as 1900. (See accompanying photo.)

The kitchen was the focus of attention in any camp. That's where the good food was. The cook had an assistant, usually called "cook-ee," or second cook, who helped with the baking, dish washing and cleaning. But cook was boss from one corner of the mess hall to the kitchen and all who walked into his domain had better take off his hat. There was no talking at

Location of this interior of bunk camp uncertain. Picture is held by Rusk County Historical Society of Ladysmith. Dating on it is also uncertain, but style of bunk shown here was uncommon after 1880 in Wisconsin. Men achieved maximum comfort when they lay with their feet towards stove (seen at right). Interior of bunk camp suggests a small crew, since most crews before 1880 did not exceed more than ten or fifteen men.

A logging camp at Altoona about 1885. This was probably a Sunday and men and horses have come out to be photographed. At left is a "Canadian peaker," six matched logs selected for the occasion. Logging sleighs in foreground. Horses at right wear housers over hames with tugs wrapped in sheepskin to protect rump of horse when backing up, usually on "cross-haul" loading. Building at right seems to be sleeping shanty, cook shanty in center with white-aproned cooks, and at far left, the stable. No two logging camps were alike; all were built according to circumstances. This one was probably for the Eau Claire Lumber Company.

table, and no one said grace although pious Catholics, who were new in camp, might cross themselves. The "Irishers," who were probably beyond grace, spoke a brand of English only slightly more intelligible than the European immigrants among them, and it was their privilege to teach the newcomers the most suitable profanity to use to inspire a team of horses.

The crews in the mid-19th Century camp ate mostly beans, biscuits, and salt pork and drank tea. By the 1880s, the food in camp had improved in variety and in taste, and by the 1890s, camp food was the envy of passing jewelry salesmen. The woods boss found that good food increased production, cut illness and improved morale.

Most camps had a "bull-cook", also called "chore-boy" who was responsible for keeping the wood boxes filled, the water pails filled, the reservoir in the kitchen stove filled, and the bunk house swept. The chore-boy could be a youth or he could be an old lumberjack with three toes missing as a result of a bad ax swing, or a twisted knee cap, the result of a log which rolled over him on the landing.

The men washed at a sink at one end of the sleeping shanty, or, in spring, at wash bowls outside. Camp towels were used in common and some men brought their own. No one ever took a bath unless he went home for Christmas, and no one carried a toothbrush.

Cook poses for picture holding knife and wet stone to sharpen knife. Man seated may be Frank Cadotte, a descendent of Michael Cadotte, Jr. of Madeline Island history. Kitchen stove at right was typical of last decades of 19th Century and was used in homes as well as camps. Black hole at far left may have been fire place. Tables are set, pie, bread and cake are on, with beef stew probably coming up.

The outhouse was a log over a slit trench, sometimes with a roof on top. A lantern hung near the door at night for anyone who had to travel as far as the slit trench. Toilet tissue consisted of old newspapers or dried leaves.

Sundays in camp were reserved for washing clothes and sharpening axes. Big iron kettles filled with water hung over a fire outside the sleeping shanty. Homemade soap was used to boil the clothing not only to remove the dirt, but to kill the lice and bedbugs which infested most camps.

Not all the men spent their weekends in camp. It all depended on how far it was to town, or how easy it was to get a ride, for prostitution was widespread in the sawdust towns. One entrepeneur with two women set up a tent close to a crew laying track for a railroad spur but was driven off by a righteous foreman.

On February 9, 1889, the Phillips *Times* carried this brief story:

> At Eau Claire last week, Andy (name withheld) and wife were sent to States Prison short term for keeping a bawdy house. It is claimed that this is the first Wisconsin case in which man and wife have been convicted together for this offense, but this is a mistake. County Judge E.W. Murry sentenced Chas—and wife to a year each at Waupun last July for keeping a dive near Phillips.

Gambling in camp was frowned upon, and a poker game higher than penny ante was forbidden. There was also a hard and fast rule about having liquor in camp, but cook usually managed to have a bottle squirreled away somewhere, strictly for "medicinal purposes" and to keep the snow snakes away from the door.

It was a policy of most logging camp foremen in the 19th Century not to have a thermometer in camp for the men to study in the morning. What they did not know, they might not feel. But, by the mid-1920s, newspapers were more common in camp and radio was being heard. William Liebert, who worked for Gagen Lumber Company in 1928, recalls one winter morning the temperature dropped to thirty-seven degrees below zero. The crew went into the woods but when the foreman arrived he realized that it was too cold for anyone to be outside and ordered the men back to camp. The horses suffered more than the crew because they were often left standing while the men worked and kept warm.

The woods boss had a room to himself in camp, either in the sleeping shanty, or in a lean-to off one of the buildings. The bookkeeper, who often shared his quarters, kept the books, the days of work performed by each man, wages coming, and wages to be deducted. He also conducted a small store in his office where the men could purchase daily necessities such as tobacco, mittens, socks and sewing materials. This store was often called a "wanigan" although a cook boat on the river was also a wanigan—anything to do with food or supplies.

For the horses or oxen there was a stable or "hovel," a term which suggests a log building. After larger camps were built and more horses were used, the stable was more often called a barn and the man in charge was "barn boss."

In addition, there was often another small building nearby where the sawfiler kept the crosscut saws sharpened, and still another building for the blacksmith where camp equipment was repaired and where the horses were shoed, and where the oxen were trussed up with a big leather cinch around their belly to lift two feet off the ground since the oxen were not able to balance themselves on three feet as well as a horse.

Some camps were so large they were more like a small village in the wilderness. One of these was Flambeau Farm at the mouth of the Flambeau on the Chippewa, and the other was the camp of Chippewa Lumber & Boom at Little Falls dam. (See sketch on page 158).

There was little illness in camp. The good food, fresh air and hard work kept the men in top condition.

Sunday afternoon in camp for Hines Lumber Company on Clover Spur off the Omaha Railraod between Winter and Draper, 1928.

Camp of James Mitchell on north fork of Flambeau River. Left, Henry Munich, William Moeller and John Maerk.

Crew in camp near Ashland, about 1895. None of men appear to be wearing anything on their feet but leather shoes and boots. Cooks hold "gabreel" horns to call crew to meals.

Corner of camp blacksmith shop with forge, fire irons and shovels used to stoke fire. At left, coal bin. Directly below fume hood hang four ball hammers, made by blacksmith, used to strike ice from under hooves of horses. Photo by author from Camp 5 Museum, Laona, Wisconsin.

Upper: oxen lacked stability with one leg off the ground and had to be trussed up for blacksmith to put on shoe. In lower right, different patented oxen shoes in collection of Howard Peddle (Iron River). Reading left to right in upper row first one was patented in 1887; center one in 1879, and last on right was hand made, date unknown. In bottom row, l. to r. is patent of 1882; next an 1880 patent, and last probably an example of an original shoe patented in 1863. Before that time all oxen shoes were forged by blacksmiths.

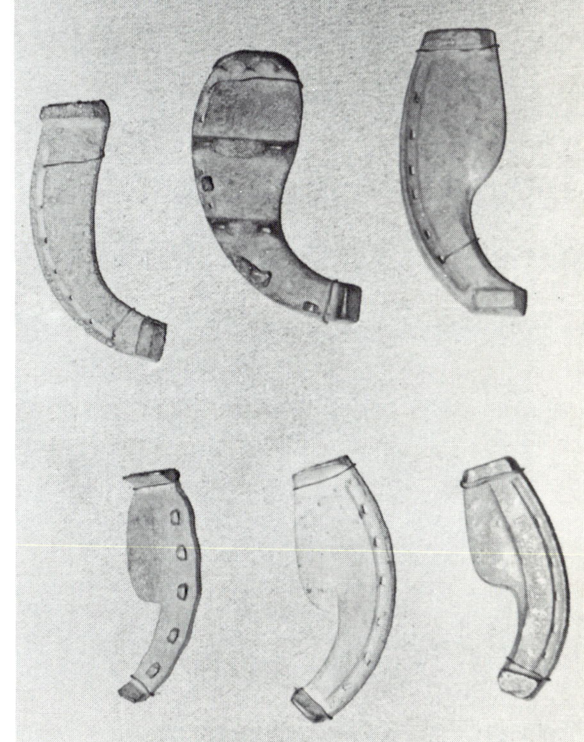

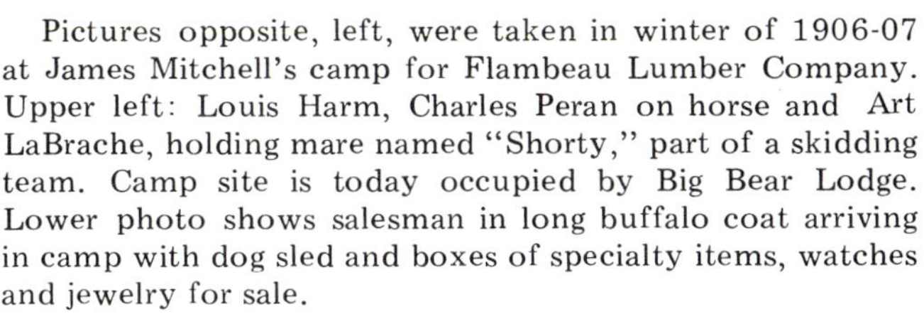

Pictures opposite, left, were taken in winter of 1906-07 at James Mitchell's camp for Flambeau Lumber Company. Upper left: Louis Harm, Charles Peran on horse and Art LaBrache, holding mare named "Shorty," part of a skidding team. Camp site is today occupied by Big Bear Lodge. Lower photo shows salesman in long buffalo coat arriving in camp with dog sled and boxes of specialty items, watches and jewelry for sale.

Camp saw filer had own shanty, worked and slept in it. He worked alone most of time but in some camps he went into the woods with portable saw vise to keep saws touched up. This one has saw in vise in front of window for best light. Rack in back holds saws, all numbered, presumably to identify teams of sawyers. Saw filer kept cat for company.

Opposite left, camp blacksmith forging horseshoe on anvil. At lower right is water barrel to cool red hot iron. In rear is forge and above, spare horseshoes. Blacksmith wore leather apron to keep sparks from burning clothing.

Tote team hauling baled hay to camp of Ole Emerson of Drywood who was logging around Twin Lakes area of Bayfield County in 1904-05. Emerson was one of last to cut virgin pine in Wisconsin.

Sunday in camp of North Wisconsin Lumber Company south of Hayward.

Location of this camp uncertain. At right, cook shanty and sleeping shanty separated by dingle. At left, blacksmith shop and a logging sleigh probably in need of repairs.

Crew of Joseph LeBoeuf, 1909, later site of village of Winter in Sawyer County.

Camp on Lost Lake (lies on the line between Rusk and Chippewa Counties) winter of 1904-05, probably for Lee Hammond.

Probably camp of Ole Emerson in Bayfield County. Man at right may be visitor, and next to him, holding measuring rule, probably the scaler. In background a mountain of baled hay next to horse barn.

The Woods

The work day in the woods of Wisconsin began before daybreak with a shout from the cook, or chore-boy, for everyone to "r-o-l-l out!" In some camps the wake-up call was "daylight in the swamp!" or "wild women in the woods, come and get it."

After a big breakfast, the men walked in the dark to their place of work to begin as soon as day broke. The men walked home in the dark, too. This twelve-hour schedule was followed in Wisconsin throughout most of the logging era before mechanization was introduced, about 1940. William Liebert of rural Marion, who worked in a camp west of Monico in 1928, said the crew he was with followed this schedule.

The object of the long day was to cut down as many trees as axes could chop, or crosscut saws could saw. Until the 1890s, trees, big and small, were chopped down by a team of two men standing opposite to each other in front of a tree, taking turns wacking at it with axes to make an initial wedge on one side. After the tree was wedged, the choppers moved to the other side and chopped a big "V" into the tree slightly higher than the wedge. In this way they knew where the tree was supposed to fall on the wedge side. The choppers had to be wise in the ways of falling a tree; it was not supposed to fall against another tree and get hung up, or to fall into a pot hole or a river, and it definitely was not supposed to fall on the crew or horses.

It has not been possible to pinpoint when choppers were replaced by two men with a crosscut saw to saw the trees down. The *Northwestern Lumberman*, a trade journal, in 1876, reported that "operators are beginning to saw down their timber instead of chopping it down with an axe. . ." In November that same year, *Mississippi Valley Lumberman*, another trade journal, predicted that the system of sawing down trees would quickly terminate the "high prices that first-class choppers have heretofore realized."

Both these reports were premature. There were crosscut saws in use before 1876, but none that effectively removed the sawdust from the kerf. Julius Kebabian, national authority on tools, believes, on the basis of xerographs from the 1876 Disston catalog, that the raker, or what was then called the "cleaner" teeth, had not been in use much before 1876, since the advertisement for this crosscut saw in the catalog states that it was only for "some parts of the country." Down to the early 1890s, the chopper was still king of the woods.

Once a tree was chopped down, or sawed down, a team of two sawyers with crosscut saws sawed, or chopped the limbs off, and then sawed the tree into logs usually sixteen feet in length. At times the swamper might help with limbing the tree, although his main job was to brush out roads for the teamster to bring out the logs. After the crosscut saw completely replaced the choppers, the basic work detail in the Wisconsin woods, roughly 1895 to 1940, was four men, namely, two sawyers, a swamper and a skidding teamster. The swamper was the low man on the totem pole, but many successful lumbermen got their start this way.

Most pine trees had three logs sixteen feet in length. A fourth log could have been made of the top in many cases, but there were too many branches to cut off and knots in the wood, and as a result, the top log was considered a cull and left in the woods to rot or catch fire.

In a lawsuit heard in 1882 between Chippewa Logging Company and a timber owner, a contract between the two parties specified that no log was to be cut less than ten inches in diameter at the top.

Sawing a tree trunk into lengths on the ground was called "bucking." After the logs were bucked up, they were "skidded" out on a road which was usually wide enough only for a team of horses or oxen. (Logs were never "snaked" out or "pulled" out.)

At a skidway the logs were rolled into "skidways of logs" and from here they were "sleigh-hauled," that is, hauled by sleigh behind a team of horses, or two teams of horses, to a "landing" which could either be on a river bank, or on a railroad siding, or near a sawmill. If the landing was on a river, the logs were decked (not piled) at least thirty feet high into "rollways." In spring the logs were released into the river and driven (floated) downriver to a sawmill.

Hillside covered with pine on Spider Lake (Bayfield County) soon to be cut for Ole Emerson who sits in sleigh rig.

Logs hauled to a railway siding were usually loaded on flat cars, or on a smaller type of flat car called a "Russel" car. From the siding the logs were shipped to the company sawmill. But most logs in the Chippewa Valley were "banked" on the embankments or rivers, or on the river ice, and from here driven downstream.

The teamster who drove the horses and skidded the logs out of the woods on a narrow road to a skidway, was often assisted by the swamper to hook the chain around the log to be skidded out. At other times the teamster might use a swamp hook, or a skidding tong, a tool resembling an ice tong. When oxen were used, there was a special man, called a "chainer," who assisted the ox teamster.

The stature of horses in the 1860s and 1870s was not large enough for heavy work in the woods. This did not mean that two oxen could skid a log out either; it usually took at least three yoke of oxen and sometimes four. The oxen were preferred in the early days because they were more steady on their feet and not as prone to accidents as the lighter horses of this period. By 1885 a larger breed of horse was introduced to the logging woods and bigger horses could move much faster on the road to the skidways than the oxen could. After 1900 oxen were seldom seen in the woods although pictures in this present volume include two or three camps where oxen were being used until 1910.

A rule of thumb in the woods held that any teamster who had more than six miles to haul a load of logs to a landing was losing money for his employer. Six miles one way meant twelve miles round trip. A good team of horses could average four miles an hour but not with a load of logs. Thus, the landings or the railway sidings had to be spaced with these limitations in mind.

When it was possible to skid a log directly from woods to a landing with a skidding tong, or with go-devil, it was known in some parts of the Pinery as "hot logging."

On the landing another crew of several men were working. These men used a "jammer" to deck the logs into huge skidways of logs called rollways. After 1900 a steam jammer was usually used in decking rollways, or in loading operations at the railway siding. A portable woods jammer, also known as a side jammer, or gin pole jammer, or horse jammer, was used in loading flat cars, too, but no matter what they were called, all used a team of horses to pull the cable which lifted the log (or logs) into the air and onto the car or sleigh.

Methods used in the woods in loading logs onto a sleigh differed according to circumstances. Sometimes it was easier to load by the simple method of "chain loading," also known as "loading cross-haul with single chain." In fact, this is the only way anything was loaded until the development of the jammers. But no one, it seems has been able to pinpoint the origin of the jammer or even the derivation of the word. The steam derrick was used in railroad construction in the 1880s and later, and probably this principle was adapted to lifting logs onto a flat car, just as ties and rails were being hoisted by the gandy dancers. One of the earliest steam jammers was developed for commercial use after 1900 by the Clyde Iron Works at Duluth, called the "McGiffert loader," It was also referred to in some picture captions as a "hoisting machine."

When loading railway flat cars the crew usually included a top-loader who remained on top of the load to guide the logs into place when they were lifted into the air. Two men on the ground set the hooks into the log, one on each end, called "pup-hookers," and two men "tailed down", that is, rolled the logs with their canthooks from the landing up to the loading zone. The two men hooking pups each held a line to the hooks, called a "sucker line," and when the log was lifted into the air and hovered over the sleigh load, or flat car, the top-loader yelled, and the two men jerked the lines which released the hooks and dropped the log atop the load where the top-loader was guiding it with his canthook.

To repeat, loading logs could be done with a team of horses and a jammer, or with a steam jammer, or with single chain.

In the 1870s a blacksmith, Alvin Lombard of Maine, began to experiment with cleated tracks to be used as wheels to propel a steam engine forward across snow-covered ground. Combining the cleated tracks with a geared steam engine made in

Here photographer has found two large pines surrounded by smaller pines. Picture was allegedly taken between Chippewa Falls and Long Lake in 1880s.

C.A. Zimmerman, who had photographic studio in St. Paul, took this picture of two choppers, probably in 1880. Both are using single bit ax. A decade later double-bit ax complimented single bit in logging woods.

Lima, Ohio, he finally produced the first Lombard steam hauler which could pull more than one load of logs in tandem. In the next two decades or so at least eighty-three units were built at the Lombard Machine Works at Waterville, Maine.

There are legends about Lombard steam haulers being used in northern Wisconsin. Collectors have tried for years to find one, perhaps sunk in a swamp somewhere, but with no success. Nor have any pictures been found. Yet, there is fairly strong evidence that at least one Lombard reached Wisconsin. On January 12, 1889, the Phillips *Times*, a weekly, ran this brief story—no headline—in its local news columns:

> The Jump River Lumber Company have a steam logger to work in Prentice. It is pronounced a success by those who have seen it. It will haul eight sleighs and will probably run about four miles per hour. They [Jump River Lumber Company] expect to put in about 18,000,000 feet during the winter.

The name given this engine is "steam logger." It was called by other names in the Wisconsin Pinery, but the most common was "steam hauler."

In 1903, the Phoenix Manufacturing Company of Eau Claire, Wisconsin, took over the Lombard patent and began to manufacture an improved model which, in a few years, could pull twice as many sleighs and larger loads than the old Lombard. The trade name for the engine was "Phoenix."

This is reproduction, actual size, of round picture frame taken with Eastman Kodak probably by Henry Stout of Knapp, Stout & Company, Menomonie. Picture, taken January 19, 1892 at "Cavick's camp" (not identified), may be earliest photographic evidence in Wisconsin of a tree being sawed down, not chopped down with an ax.

To many people the logging woods was a romantic place, filled with men of great strength and tales of great length. *Harper's Weekly*, on March 28, 1885, ran a brief story called "The Wisconsin Loggers" and since it was written on the basis of a visit by a correspondent to Fifield, it falls within the purview of this book and it reads as follows

> The big timber belt of northern Wisconsin is a region full of charm. The men who do the logging are Spartans in one way, but not in another. They eat well and they drink well. They tire themselves with working at the logs, exhilarate themselves by the pursuit of game, and regale themselves as well as men do in the effete communities. City men, accustomed to ease and luxuries, love to get among them. The ease to which the men of the town are used is not rudely broken in the Wisconsin woods. They refrain from taking a hand at the logs, and it would be better to say that they comfortably wait for game than that they vicariously pursue it. One of the sections of Mr. Graham's picture shows a deer-box, built upon a tree, and with decent steps leading up to it. These boxes are placed in runways, which are everywhere, one of the most celebrated crossing the railroad track. It is no hardship to sit in a deer-box and shoot a deer when he comes along. And then there is the varied feeding and the comprehensive drinking, quite famous in these primeval woods.

The town of Fifield, on the Wisconsin Central Railroad, is the centre of the logging district. Go a hundred miles to the east or to the west of it and you will not find a village or a town—nothing but woods. You know when you get to Fifield that you have reached a logging district. The station platform, the floors of houses, everything that is under foot is full of holes. They are made by the spikes worn in the boot soles of the loggers. All the flooring in logging towns is rapidly spiked to shreds, and must be frequently renewed.

The huts of the loggers are built of logs, and are very warm and comfortable. The horses are of the best breed. The iron runners on the sledges are polished and smooth as glass. Perhaps the hardest work the loggers have to do comes when a jam occurs on the Chippewa River. To loosen up a jam is a job of log-rolling that might baffle a Legislature [politician].

The *Harper's* correspondent did not spend much time in one of the "huts" he refers to or he would have known that they were not comfortable; they were bearable. Nor were they called "huts." This was a term heard in the cut-over lands after the loggers had left but not applied to a logging camp. As to hunting from "deer-boxes," that was a luxury enjoyed by a few. Most working men could

After tree was chopped, or sawed down, two sawyers, working as team, cut off limbs and "bucked" up log into lengths usually of sixteen feet. Note the top of stump at lower right. Tree was an isolated giant. Other trees in distance are mostly hardwoods.

This picture dated 1911. Although horses had largely replaced oxen by that time, some camps were still using both oxen and horses for skidding. Picture was taken in camp of Jack Ryan in Sawyer County. Ryan was woods boss for Chippewa Lumber & Boom in early 1900s and refused to pay a bill submitted by John F. Dietz for serving as watchman at Price Dam on Brunet River, a feud which was carried over to Cameron Dam on Thornapple River.

not afford a rifle. Finally, the calked boots were worn not by lumberjacks in the woods, but by river drivers, in this instance probably by men working the drives on the south fork of the Flambeau, not the Chippewa.

Meanwhile, if one were to study a map of the Chippewa Valley, it might be assumed, when lumbering began, that operations advanced north along an east-west line stretching across Chippewa County. But this was not the case. Instead, camps were located deep in the forest all over the county since it was here and there that a particular lumber company had acquired pine stumpage. The plat of 1888 of Chippewa County, for example, reveals the location of timber forties purchased by the leading lumber companies; it was to these forties, no matter where they were located, that the woods bosses sent their crews in to begin cutting, skidding, hauling and decking.

Between the winter of 1888 and 1894 logging operations in the Chippewa Valley reached their zenith. Some idea of the extent of these operations may be judged from a report appearing in the Chippewa *Times* and copied by the Phillips *Times*, December 22, 1888, which covers the anticipated winter cut for Chippewa Logging Company and Mississippi River Logging Company. (This must have been a short time after the Chippewa Logging Company was incorporated). Here are the general locations of camps and the estimated cut, in board feet, for the winter of 1888-89:

Main Chippewa	75,000,000
West Fork	45,000,000
East Fork	10,000,000
Main Flambeau	13,000,000
South Fork	38,000,000
North Fork	31,000,000
Little Chief	15,000,000

An average log could be skidded out of woods to logging road with skidding tongs over snow-covered ground, but log, seen here, was too big for two horses. It was rolled onto go-devil or dray which lifted butt end off ground. Picture taken about 1890.

Main Jump	7,000,000
South Fork Jump	13,000,000
North Fork Jump	5,000,000
Thornapple	12,000,000
Couderay	30,000,000
Brunet	5,000,000
Nail Creek	4,000,000
Skinner Creek	9,000,000
Fisher River	5,000,000
Wiergor	5,000,000
Elk River	10,000,000

In addition to these figures, Chippewa Lumber & Boom was expected to cut 65,000,000 feet.

Most of the logs cut for the big lumber companies in the Chippewa Valley were cut by contractors, also known as jobbers. The contractor built his own camp in the woods, hired his own crew, and brought his own horses or oxen and sleighs. His contract usually called for the delivery of a specific number of board feet of logs. The number of contractors who drew their supplies from just one railway point in the winter of 1889 is revealed in a story appearing in the Phillips *Times* on February 2nd. According to this, there were thirty camps drawing their supplies from Phillips, most of them, no doubt, located in Price County along Elk River and the Flambeau. Here are the names of the contractors, or companies, the number of camps and the number of board feet they were expected to bank that winter, that is, put on a river bank in rollways:

	camps	board feet
Quail Bros.	2	10,000,000
Alex Malcomb	2	12,000,000
Richard Medley	1	3,000,000
Thomas Meredith	1	7,000,000
William Farrell	1	1,000,000
Allen Jackson	1	1,500,000
Dells Lumber Co.	2	10,000,000
Bruno Vinette	1	5,000,000
Pat McLaun	1	3,000,000
Anderson Bros.	1	9,000,000
Joe Elourneau	1	3,000,000
Badger State Lbr. Co.	1	6,000,000
Hatton & Fordyce	2	3,000,000
Wellington McMullen	1	1,000,000
J.J. Kennedy	1	3,500,00
Forbes Bros.	1	3,000,000
Phillips Lbr. Co.	4	12,000,000

On February 16, 1889 the Phillips *Times* listed the names of contractors and companies drawing their supplies from Fifield with number of camps and board feet to be banked that winter. The spellings follow the original news report submitted by Kate Egnon:

	camps	board feet
Lee Hammond	3	13,000,000
Wm. Fawlds	3	10,000,000
J. Sweeny (part)	1	10,000,000
P.S. Davidson	2	8,000,000

This appears to be usual method of loading logs with single chain. But in this instance, loaders have improvised by tying block (visible with magnifying glass) to tree and running chain through it to evener behind horses. Evener is suspended in air above rump of horses to hold chain and log in mid-air for benefit of cameraman. Men on ground at right called "send-up" men; top loader stands on top, teamster drives cross-haul team, and scaler holds measuring stick.

McDonald & Gilbert	2	8,000,000
Geo. Cochran	3	7,000,000
D.L. Sugrue	3	7,000,000
F.D. Lindsay	2	6,000,000
R. McMurdo	1	6,000,000
Poloquin & Cornezes	1	5,000,000
A. Smith	2	5,000,000
Felix Blodd	2	5,000,000
Stickney & McPherson	1	5,000,000
John Pearl	1	5,000,000
P.H. Leonard & Co.	2	5,000,000
P. Traux	1	4,000,000
J. Bertels	1	4,000,000
Galloway & Harris	1	3,500,000
Medley & Connors	1	3,500,000
Robert & Laterns	1	3,500,000
Corn & Dukir	1	3,000,000
Poole & Reed	1	3,000,000
D. Dyer	1	3,000,000
Wilcox & McKenzie	1	3,000,000
Ed Flanders	1	2,500,000
F. Bradley	1	2,500,000
Pooier & Lalles	1	2,500,000
T. Sugars	1	2,500,000
John Medley	1	2,000,000
J. LeVally	1	2,000,000
Singleton & Turner	1	1,500,000
John Medley	1	1,500,000
J. Murray	1	1,500,000
John Avard	1	1,000,000
Moses Brown	1	1,000,000
F. Rivard	1	1,000,000
Geo. Deripple	1	1,000,000
H. Albright	1	1,000,000
J. Thevarge	1	400,000
A. Patterson	?	350,000
B. Hilgart	?	300,000
J. Labermeier	?	300,000

To cut as many trees as possible in one day depended much upon a steady supply of manpower. Many mill owners used the contract as the most successful way of achieving this goal. For example, Knapp, Stout & Company employed the contract system as early as 1868. In November 1872, an ox teamster and crosscut sawyer signed a contract to work four and a half months at $20 per month. Payment was made at the "end of time" with $4 deducted per month if the worker left the woods early.

According to the testimony of Dale A. Peterson, who did his Ph.D. dissertation on lumbering in the Chippewa Valley, Knapp, Stout & Company once discharged a man because he not only disobeyed orders but encouraged others to follow suit. The worker brought suit against the company and the court found in favor of the plaintiff. Yet, Knapp, Stout & Company continued to use what was known as the dockage system. In the 1880s a state factory inspector made this comment: "Peter Larson got killed in the woods while in the company [Knapp, Stout] employ. His wages were docked 20 percent for not working his time out. His widow returned to Norway. I can't blame her."

Wages in the logging woods varied from river to river, but some idea of the general level of wages in the fall of 1889 appears in the Phillips *Times*, November 17, when it is learned that the Lumber & Log Owners Association of Ashland had adopted the following wage scale for the winter of 1889-90:

Cooks, from $40 to $45 per month.

Cookees, $18 to $22 per month.

Four horse teamsters, $32 to $35 per month.

Two horse teamsters, $24 to $28 per month.

Ox teamsters, $26 to $28 per month.

Loaders, $28 to $30 per month.

Head sawyers [that is crosscut sawyers bucking up trees], $22 to $24 per month.

Second sawyers, $20 to $22 per month.

Choppers, $26 to $28 per month.

Road workers [swampers], $15 to $20 per month.

Blacksmiths, $35 to $40 per month.

Blacksmith and woodworkers, $40 to $45 per month.

Wood tinkers, $30 to $35 per month.

Cooks, blacksmiths and tinkers were to be employed by the calendar month with no allowance for extra time. Time checks would be paid on May 1st, i.e. the next spring.

From this it seems that none of the lumberjacks in the woods got paid until they were about ready to leave the woods, unless they were discharged on short notice or "sent down." This wage schedule also makes clear that in the Ashland district, at least, trees were still being chopped down, not sawed down.

Drawing of an ideal ground plan for logging operation by Randall E. Rohe. Logs were cut between rows and brought out to skidways where they were later loaded on sleighs for haul to landing on river or railway siding. But rough terrain usually did not permit a system of logging roads as logical as this.

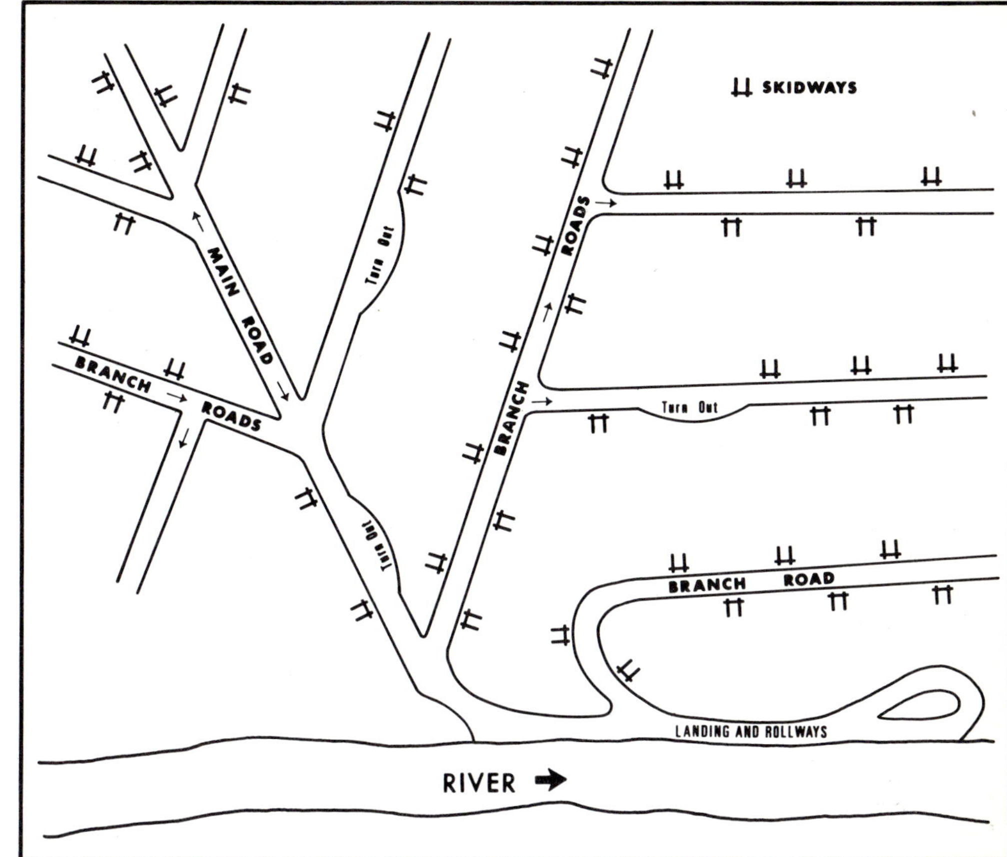

Donald Lockery holds canthook to show how it is used to roll a log forward (left) or backward (right). Pictures were taken in author's woods, Portage County, November, 1981.

Above, two sawyers prepare to saw down a big white pine. Long stick in ground at left was used to measure log lengths after fallen tree was being bucked-up. Men are wearing rubber boots which suggests picture was taken after 1900. In background are smaller trees, rampikes and slashing.

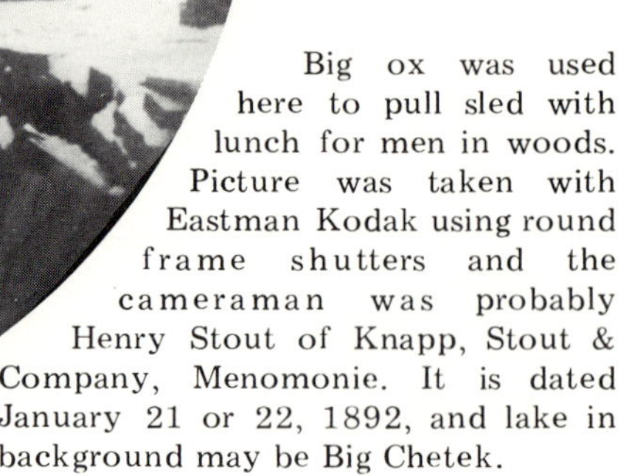

Big ox was used here to pull sled with lunch for men in woods. Picture was taken with Eastman Kodak using round frame shutters and the cameraman was probably Henry Stout of Knapp, Stout & Company, Menomonie. It is dated January 21 or 22, 1892, and lake in background may be Big Chetek.

Right: Sled with buckets of hot food arrives in woods for logging crew, another picture probably taken by Henry Stout in 1892.

Below: J.C. Amundson of Ladysmith took this picture. Horses wear housers over hames to prevent snow from working into collar and causing sore shoulders.

Ox teamsters at right have skidded these pine logs to skidway. Logs rest on long stringers, the ends farthest away from the tote road higher so that when loading operations begin, logs may be more easily rolled or "tailed down" towards sleigh. Picture was probably taken for Eau Claire Lumber Company in a camp at Altoona in the 1880s.

Franklyn L. Stevens, woods boss on Amaco Lake in 1893. Blocks between first and second tiers of logs on sleigh called "grousers," were inserted to steady load until chains could be tightened. Stevens wears leather boots which were still popular with loggers and river men into the late 1890s, when rubber boots began to replace them.

Below: crew for Ole Emerson, probably on Spider Lake. Picture shows use of tong to skid log out of woods, an innovation of late 1880s in Michigan. Two youths at left are probably swampers who brushed out trails for teamster to skid logs out to road. Man in center holds canthook.

In some camps, saw filer went into the woods with crew to keep crosscut saws sharp. Here portable wooden vise is fixed to stump. Hammer in front was used for swaging raker teeth. In his left hand filer holds a gauge which he uses to get teeth filed to exact dimension. Type of crosscut saw shown here was probably used more than any other, that is, two cutting teeth alternating with one raker. The raker removed the sawdust in the kerf.

Right, an icing tanker used for sprinkling water on logging roads to create ice for sleigh tracks. Water was raised from hole in pond or river, one barrel at a time, by horse power, the barrel rising to top of tank on skid poles and tripped.

▶

Below: a rutter for Rice Lake Lumber Company, 1909. This was a sled with two heavy runners made by camp blacksmith or carpenter and used to create ruts in snow for sleighs to follow after ruts had been iced by tanker seen at right.

This man was called a "road monkey" or "hayman-on-the-hill." It was his duty to put hay on sleigh tracks for logging sleighs going down hill, and to keep snow in tracks on uphill. He usually had to work at night and therefore carried a lantern. Boot packs and coat style suggest picture was taken in 1920s.

Left: tracks show road monkey has laced rut at left with hay to slow down load coming downhill.

Below: sixteen horse team hauling Norway Pine cut to lengths of thirty-two and thirty-four feet, probably for telephone poles. Load is coming up grade off lake for Lammer Brothers Lumber Company of Drummond, 1902.

Photograph, dated February 16, 1897, was taken by O.G. Merriman of Eau Claire at camp of I.K. Keer near Phillips, Wisconsin. It shows how big load of logs could be decked by traditional method of "loading cross-haul with single chain." Skid poles on which log rises can be seen at right leaning against top tier of logs. At far left stands cross-haul team. This was "star" load decked especially for benefit of photographer, not ordinarily seen in woods. Nor will two horses budge it. Bottom tier of logs held in place by corner binds, but others have wrappers, i.e. chains wrapped around. Figure at lower right gives number of board feet in load. It has been thought that load this high could not be decked without jammer, but picture proves it could be done with single chain.

Some time between 1893 and 1905, when above picture was taken, a new technique was developed for loading logs on sleighs or decking logs onto rollways, namely, with steam hoist, a form of derrick, or horse-powered hoist better known as the woods jammer, or side-jammer, or horse jammer. Some were made with a single jin pole. The one shown above is an A-frame. Jammer rests on framework of small logs or stringers which rest on sleigh runners. Later models were made so that framework served as sled. By sharpening ends of logs like skiis, jammer could be skidded around forest by team of horses. Jammer seen above was being used by crew of Ole Emerson, in 1905, probably Spider Lake.

A steam hoist or jammer operating from railroad flat car. Picture was taken by Charles B. Lewis, an Eau Claire photographer.

Loading logs from skidway to railway cars with steam jammer on Hammond spur near Birchwood (Barron County), 1906.

Loading flat car on Omaha line at Clover spur between Winter and Draper for (Elmer) Larson Logging Company, 1928. Loading is being done with swing-boom jammer, and two men have climbed on log for benefit of cameraman. Long car stakes were standard for loading by this time. At left, tailing down, is Henry Munich.

A star load at Lost Lake for E.S. ("Lee") Hammond who cut forty million feet of logs in winter of 1908-09. His woods boss was Thomas Boury, and top-loader, George Villiard of Radisson. Ballie, Nellie, Randy and Dan, the four horses, allegedly pulled this load three miles. It was nineteen feet wide and twenty-four feet high and scaled 21,770 feet. Leaning, one hand on white horse, is Herman Sidenkrons.

Axle pin for dry roll.—Michael F. Slasinski collection.

◀

This is an average load of logs for two teams of horses. It was taken at Camp 2 on Blueberry Lake on the Court Oreilles Indian Reservation for Bekkedahl Lumber Company of Couderay in 1916. Small logs suggest these may be second growth.

Caption above appears on original picture. John Colbroth's Camp I was located south of Hayward. He was contractor for North Wisconsin Lumber Company's big mill in Hayward. Colbroth Lake, west of Grindstone, was named for him. Logs in picture were probably loaded with single chain.

This "Canadian peaker," also called "Michigan peaker," was probably decked with jammer at camp near Park Falls in 1921. Scaling 20,040 feet. With exception of one or two logs, others appear to be of uniform size, all selected for picture. One log held in air by two chains has been hoisted for effect; it is not going anywhere because there is no room at top. Two horses could not pull this load, nor could four. Slightest dip in road would send it flying.

A mountain of Norway pine, not seen in Wisconsin after 1920, but in Upper Michigan where this was taken. Martin McGinty, scaler, holds rule stick to log to determine board feet. But picture is posed. Scaler usually measured logs at landing as they were hauled in, or before they left woods. Photo taken for White River Timber Company of Bergland, Michigan.

Above: Most pike poles were three meters long (about fifteen feet) and had single spike on business end for pushing logs in water. This pike pole has hook and bent spike and was used occasionally in hot pond to push logs, but mostly for pushing or pulling ice blocks for local ice house. Larger logging camps had ice box and put up own ice.

Sawmill overlooking Chippewa River at Ella, early 1900s. Mill is sawing four foot bolts.

In the winter of 1908, W.E. Moses and Thomas Gaynor contracted with Chippewa Lumber & Boom Company to haul company logs stranded on Thornapple River over to Babbs Island on north fork of the Flambeau (south of Big Bear Lodge). These logs had been lying in dried-up flowage of Cameron Dam after dam washed out in 1906. Some of the logs lay on John F. Dietz' land, but majority on company land, and deal was struck to give Dietz all logs lying on his land in exchange for promise not to harrass anyone moving other logs. Moses made deal for C.L.&B. Logs were hauled with this Phoenix steam hauler manufactured in Eau Claire.

Above: Steam hauler for Fountain-Campbell Lumber Company which had camps east of Cornell and later near Ladysmith. Below: Rice Lake Lumber Company's steam hauler on Leonard Spur, east of Seeley in Bayfield County, 1907.

Pictures, above and below, were taken at Rice Lake Lumber Company camp on Stout spur near Draper (Sawyer County), winter of 1910-11. Above, general view of rollways being decked with steam jammer, engine enclosed in shed. Lower: close-up of decking technique using double chain, probably on cross-haul, not with jammer. Logs are hemlock and many have punk rot. In picture above note idle icing tanker at right.

Above: decking logs with crotch chain and jammer for Lee Hammond camp near Winter, about 1910. Cross-haul team stands at right. Two men next to skid poles hold "sucker lines" attached to pup-hooks at each end of log. When log reaches top of deck, sucker lines will be yanked to release log where top-loader wants it. Logs here appear to be yellow birch. Below, hemlock logs decked at landing of New Dells Lumber Company of Eau Claire near Kennedy, winter of 1913-14.

The steam skidder was developed in Michigan in the 1880s and brought to Wisconsin, but used only in two or three camps. The one shown here was for Shell Lake Lumber Company. By the use of these complicated cables, logs were skidded, that is, lifted directly out of woods where they were cut onto flat car shown here to left of engine shed. Head spar looms over shed and heavy cable above is connected to tail spar some distance to left of picture. By use of traveling block, cable went back and forth carrying logs to flat car. This technique was adopted widely in the Pacific Northwest, but in Wisconsin it held little attraction. It took just as much time to bring out a small log as a large one, and, in most cases, logs were too far from railway spur.

Several loads of logs approach landing in Ole Emerson camp near Cable, 1904. Loads are wide and roads were probably iced. In foreground is close-up of logging sleigh. Bunks are turned sideways but angle reveals how bunks were balanced in center of beam below. Holes at outside for sway bars which helped steady load. Chains running through second holes are for the corner bind, round hooks showing, but not fid hooks.

Canthook, red oak log, and go-devil.

Picture of an icing tanker probably taken by Henry Stout of Menomonie, Wisconsin. Photographer was using early Kodak made by Eastman which took round frame. This picture, about twice enlarged, was taken on January 21, 1892 at camp on Red Cedar river where C. Johnson was woods boss. This may be earliest photograph of an icing tanker presently available in Wisconsin. Instead of lifting water barrel on cross-haul in the manner of a later generation of teamsters, this one employs the lever. Weight at left balances weight of barrel and it seems that the barrel is lifted over the end of tanker to be dumped. Hole in ice where the barrel was dipped not visible. Two men at left are probably company officers accompanying the photographer. No lumberjack would wear a white scarf around his neck.

The Drive

It all seemed so easy. All the lumbermen in northern Wisconsin had to do was to get their logs to the bank of a river, wait for spring and then roll them into the water. Even creeks were made navigable by building small dams to raise the stage of water. Thus, the rivers were both highways and railways and without them, the pine harvest on the Chippewa might have been held up for another generation.

After the logs were rolled into the river in spring, they were taken in hand by "drivers," that is, men who guided the loose floating logs downstream. Following the logs downstream was called a "drive." The men, in a sense, were driving the logs before them and it was their business to see that they kept as many of the logs as possible in front of them and not behind them. But no matter how hard they might try, some logs floated ashore or got caught in sloughs, on stones or sandbars. It did not take much; one log attracted another and soon there were many more and if they were not freed at once, a jam would form which would block the entire river. The jam or "jam-up," was the dread of all drivers. It could delay delivery of a raft of logs for days and even weeks and it often blocked the flow of the river itself, raising the water above the jam higher and higher, and lifting incoming logs atop the jam. In small jams the river often went around the jam, washing out the banks on one side like a beaver house will do in the middle of a trout stream.

The drivers were nicknamed "river pigs" but most of the time they were called "drivers" or "peavey men," that is, men who handled the peavey canthook.

Considerable planning went into a log drive. Men had to be hauled up to the point of beginning, or they walked upstream, or some were recruited locally. They had to be fed, and tents and blankets had to be provided. A floating kitchen, called a "wanigan," had to be built in addition to several *batteaux* (boats) to carry the drivers back and forth across the river. Some men carried peavies, and some long poles with spikes in the end to push the logs away from shore. The pole men worked mostly from shore; the peavey men mostly from the *batteaux* or directly from a log.

In addition to tents, there were permanent installations where the men camped, for example, the camp maintained by Chippewa Lumber & Boom Company at Cameron Dam on Thornapple River. Here there were three log buildings, a sleeping shanty, a cook shanty, and a horse barn. There always had to be accommodations for horses because tote teams were coming and going, bringing in supplies to camp and personnel from the nearest railroad town or base camp.

There were other staging areas and "stopping places" which catered to the log drivers such as Grand Rapids Hotel near the present city of Bruce in Rusk County, and Hall's and Thayer's farther north on the Chippewa. The Shaw Lumber Company maintained a big staging area at "Flambeau Farm" located opposite the mouth of the Flambeau on the Chippewa River. Here there were sleeping shanties for both transients and permanent crews. After the railroad came to Fifield, men who drove the south fork of the Flambeau were usually recruited here. On the north fork of the Flambeau, most of the drivers were recruited from or lived in Park Falls.

There were two crews within the overall crew on a drive, one the "jam" or "beat" crew, and the other the "rear crew." The forward or beat crew consisted of ten to fifteen men, the rear crew as many as thirty or forty men. Others were employed to row or pole the skiffs, known as "batteaux," and others served as cooks, lunch carriers and couriers, the latter often called "telegraphs" or "cowboys."

It might be said that the beat crew took the "point." It went ahead; it was called a beat crew because two or three men were assigned to certain stretches of the Chippewa River called "beats" to watch and guide the logs floating downstream. On the Wolf River, the forward crew was called a "jam" crew, not a beat crew, although the term "rear crew" was used on the Wolf to mean the same.

The beat crew was composed of men who had the most experience on the river. They were the most agile, the most fearless. Some were known as "whitewater" men because they could ride a log through a rapids where the water ran white. Not much of this was done but it was done when necessary.

William Hoyer, born in Chippewa Falls, last survivor of the "*Chippeway* drive," described the beat crews this way:

> On the main drive Archie Smith, our boss, would pick out fellows he knew could take care of themselves, . . was good on logs, you know. There was all kinds of little rapids, centers here and there, and he'd send two men, beat men, to watch a certain stretch of the river, not in boats, but you just took care of yourself. You rode a log to save the boat. Like me and Jack Sugars there. We were beat men and they'd have the beat men strung out, just far enough apart so they wouldn't have too far to walk back.

In the above, Hoyer refers to a "center." This was a small jam of logs caught on a rock or sandbar in the middle of the river. It was not really a jam-up. In many cases dynamite was used to attack a center. When a center was coming apart there was little time for anyone to get off. Said Hoyer:

> You'd get sixteen men in that boat, unless you were going out to a center. All of them would get out of the boat riding a log each one of them. All excepting about four if you were going out to a center. You had to pick them four guys up. You drove your pick [peavey] into a log to hold the boat there, and the rest of the guys was supposed to flatten that center out by pushing or pulling the key log loose. After they did that they had to be fast and come runnin' across the logs and jump into the boat.

The rear crew "brought up the rear," that is, after the beat crews had gone ahead with the main drive of logs, many logs were left behind, caught in sloughs and along the banks. When they formed a phalanx of logs along a stretch of the embankment, they were called a "wing." Some places these wings were allowed to stay in place to narrow the channel of the river and raise the stage of water in the center of the river. But usually, the wings were removed by the rear crew working its way downstream. Although many of the rear crew were local farmers and inexperienced woodsmen, they had to be just as rugged as the beat crew men because they were often wading in icy water up to their pockets, pushing and turning a log caught on a sandbar into the current. At other times, logs, in

Lunch on way for drivers.

high water, floated so far ashore they could not be rolled back by the rear crew with peavies. After the flood receded, a team of horses or a yoke of oxen was used to bring the logs back to the river bank. The log was not skidded in the manner used by a teamster in the woods. A rolling dog and link was

Log drive approaches Dells pond.

Loose floating logs on Chippewa, probably above Durand. Jam piers and boom at right were built to keep logs in main. current and not permit them to drift ashore. At left, a narrow road for drivers to follow when gigging back.

Breaking up a "center." D. H. Brown of Chippewa Falls took this picture at Eagle Rapids, 1909.

pounded into each end of the log and a light chain ran from the dogs to the single tree behind each horse. In this manner the log was rolled like a wheel back to the river, and, because the log was on dry land after the flood receded, it was called "dry roll." Dry rolling is mentioned several times in the journal kept at the Little Falls dam described in a later chapter.

The beat crew was assisted by "pole men," that is, men with long pike poles who walked along the shore line pushing logs back into the current if they clung to an embankment. Some river men called this the "jam crew."

The rear crew was much larger than the beat crew because it took more time to get a log which was stranded on a sandbar or in a slough back into the river. This was often called "sacking," but the term also had other applications, for example, when a rollway of logs on a river embankment was being rolled into the river in spring, some drivers called this "sacking."

There were never any two drives alike, and there were any number of variations possible, all according to circumstances, stage of water, the weather and the wind. Some driving crews took their logs all the way to the Mississippi River, others only as far as Chippewa Falls or Eau Claire.

A. A. Bish of Chippewa Falls took picture probably on last drive for Chippewa Lumber & Boom in 1909 at request of William Irvine who gave each member of crew a copy as remembrance of things past. This crew will break up "center" in background. In distance is steel bridge built in 1906 below Little Falls dam (Holcombe). Batteau at left includes Scotty McTavish (in stern), Harry Andrews, Walter Sugars, Dave Mullen, Johnny Mack, Charlie Goldy, and Charlie Ermantanger (standing in bow). Included in the second batteau on left are, l. to r., Vet Harp, Joe Violet, Johnny Burke, (next hidden), Will Lyden, Billy Smith, "Red" Hough, Hank Johnson and Walter Smith (one not identified). Included in third batteau, l. to r. are Alphonse Conor, Jim Gorman, Jack Hedrington, Charlie Corbine (a Chippewa), and Pete Harp (next two not identified). Batteau in rear includes, l. to r. "Red" Paddy McDonald, "Spotty" Dan McDonald, Indian Joe, Fred Cheeters, Roy Smith, Jimmy Smith, "Muskrat Joe" DeBault, John LeTendre, and Billy ("Boo-hoo") Hoyer, who stands in bow. Batteau at right includes Archie Smith, river boss, who stands with stick over shoulder. Seated include Charlie Browning, George Craig, and Bill McIntrye.

And there were also auxiliary crews recruited at the lower end of the Chippewa. On July 30, 1880, the Pepin County *Courier* reported that the Beef Slough Company had let contracts to two parties to put in the logs along the river which had strayed over the banks in the big flood of 1880. Mr. D. Pridle got the contract to put in the logs on the east bank of the Chippewa from Browning's Lake down to the mouth of the slough below Round Hill, and John Gilmore got the contract for the logs lying on the west bank from the head of Nine Mile Slough to Round Hill. (Nine Mile Slough lay less than ten kilometers above Durand.)

The *Courier* on November 9th, reported that John Gilmore had a small crew of drivers at work during that week "starting logs off the bars near town (and) if the water raise (sic) enough, a full crew will probably be put on and as many logs as possible put into the slough before the river closes." (John Gilmore was the father of Albert Gilmore, boom boss at Round Hill.)

Somewhere on the river at Eau Claire in 1877 there was a works where logs were being sorted according to their ownership denoted by the end marks, or stamps at the end of the logs, or the bark marks. On July 12, 1877 the Eau Claire *Free Press* reproduced a conversation which one of the workers on the sorting gap was having with himself. He was talking to the logs because, in a sense, they had acquired a personality by the fact that they belonged to someone the sorter knew by the end marks. This is the way the reporter for the *Free Press* recorded what he overheard:

> In with the Norway. Outside three times. Old Buff. I.K. two times. Girdle hatchet. Shaw the big one. Outside all three. Take the little one in. The sluice way is jammed. Buff again. Look at the stamp mark of that old soaker. Bill all right, out she goes then Beef Slough, five times. Uncle Daniel. Give this to Ingram to keep his courage good. What time is it? I'm getting hungry. Eleven o'clock and thirty minutes. Thank God for that. Down on that wedge log, she's jamming boys. All right, let 'em come. Give me a chew of that. Catch my hat—yes, and my pole, Bill. Oh, give us a rest. Beef Slough. Shove the long one along lively. That river sluice way will jam. Punch that long one around lively. Shaw, all right. They are running over now. Some jam has broke away above.

"Old Buff" was, no doubt, George Buffington, and "I.K. two times," refers to Ingram-Kennedy Lumber Company. Girdle hatchet was a nickname for an end mark. "Shaw the big one," means a big log for Daniel Shaw Lumber Company. "Bill" is probably William Carson, and "out she goes then Beef Slough five times," means the sorter had spotted five logs for the Beef Slough Company. "Uncle Daniel" again refers to the Daniel Shaw Lumber Company, and Ingram, who needed some logs to boost his courage, was Orrin H. Ingram, one of the leading lumbermen in the Chippewa Valley.

"Down on that wedge log," means the sorter is pushing a log under water to another spot before it creates a jam. And it must have been a windy day, since he is telling someone to catch his hat and he apparently lost his pole for a moment, too.

In the late 1860s when log drives were in progress along the Chippewa River, mill owners stopped or held up the logs at a large, natural storage pond at the mouth of Paint Creek a short distance above Chippewa Falls. Since the logs, presumably, had end marks or bark marks, they were sorted at a gap and diverted into corrals or pens made of connecting boom logs. These boom logs were held together by short pieces of chain attached to link dogs pounded into the ends of the logs.

If a mill owner found the marks of another mill owner or jobber on a log he was supposed to allow it to pass and not to interfere with it. But in order to sort his own logs, a mill owner usually had to get the logs of other mills out of the way. This often led to delays, some honest, some not so honest.

Lest there be any doubt on this point, J.G. Thorp ran an advertisement in the Eau Claire *Free Press* under date of March 10, 1868, advising that all logs in the booms on the Eau Claire River would be run through "with least delay possible." In fact this same advertisement continued to appear weekly all spring. But on June 25th, a news story in the same paper reported that a recent drive on the Chippewa River was a failure which would force mills to shut down. Only the Daniel Shaw Lumber Company and Wilson, Tarrent & Company would not close.

Big Falls on Flambeau River about fifteen kilometers above Ladysmith. In distance, a "center" waiting for log drivers to break it up.

The failure of the Chippewa drive was not due to a conspiracy on the part of any boom company. It happened time and again, and the reason was simple: the stage of water was too low, or, too high. The mill men were ever thus at the mercy of nature's whims.

In 1869 a jam formed at Paint Creek Rapids and when it began, it was impossible to warn the crews on the banking grounds farther upstream to hold their logs back. As a result the logs piled up to a height of several meters and backed up the river almost to Eagle Rapids. It was the worst jam in the history of the logging industry in Wisconsin. Frederick Weyerhaeuser happened to be in Chippewa Falls on business and walked out from his hotel to have a look at it, knowing full well that thousands of the logs in the jam belonged to his

sawmill at Rock Island, Illinois.

The story of the Big Jam reached the editors of *Harper's Weekly* and they sent a photographer, Nathan A. Preston of Eau Claire, to take a picture of it. The editors could not make a zinc block and print the picture, a technique developed a decade later, but they had one of their engravers make a copy of it and this appears in the *Weekly* on June 6, 1869. (See accompanying photo.) On another page of the *Weekly*, the Big Jam received three paragraphs and this is how it is described:

> We illustrate on page 360 a characteristic incident of the late freshets in Wisconsin. The piers shown in the picture were put in by Pound, Halbert & Co. to protect their works below from damage by the heavy ice freshets in the spring, and also to detain temporarily the logs, to prevent them from running into the river until they could sort out their own logs from the general drive. Some of these piers stand in forty feet of water, which is quite deep for a long distance above the piers. There was a much larger amount of logs put in during the last winter than ever before, and probably about sixty million feet were put in on the ice, so that when the latter went out this large body of logs moved together and choked up the river. Immediately behind this came the general drive of probably ninety million more. The jam formed something more than half a mile above the piers, and extended up the river more than a mile and a half—one solid mass of logs, and there they stuck, refusing to move.
>
> A large number of men were set to work to break it up, which they finally succeeded in doing, when the entire mass moved together for half a mile, presenting one of the grandest scenes ever witnessed, although attended with some danger to those who remained on the mass while it was in motion. The piers proved strong enough to stop the logs, else everything below in the shape of piers, booms, and logs would have been swept away. The logs are in many places piled twenty feet above the level of the water.

If Preston filed this story he erred in saying the Big Jam had already been broken. Whoever it was, the reporter was misled by the sudden movement of one of the "grandest scenes ever witnessed," since other sources agree that it took most of the summer to break it up. And it is also doubtful whether anyone was riding this mass of logs unless by accident. The piece in *Harper's*, nevertheless, offers a rare instance of reporting by an eastern newspaper about the "late freshets in Wisconsin."

Another log jam provided an excuse for the only strike by drivers in the Chippewa Valley. According to a letter written by Captain Charles H. Henry to Weyerhauser, the strike occurred on the south fork of the Flambeau in the spring of 1888 or 1889. A jam had developed a short distance below the Pike Lake dam and the crew decided that this would be a good time to call a strike for a raise of fifty cents a day. Captain Henry, a man who won his commission in the field during the Civil War, was one of the walking bosses for Weyerhaeuser and he had learned that a logging contractor, probably over on the north fork of the Flambeau, had just completed a drive and paid off his men. Henry broke the strike on the south fork

Worst log jam in history of Wisconsin occurred at Paint Creek Rapids, above Chippewa Falls, June 1869. *Harper's Weekly* sent photographer, Nathan A. Preston, of Eau Claire, to take picture. At right are jam piers. These were built to slow up logs, or as sorting point. Piers are built of timbers, filled with rock, and slanted on upriver side. Pressure of river held piers in place. At right, walking boom for drivers. Underneath planks are three logs, or square timbers, bound or spiked together to form flotation on which planks rest. *Harper's Weekly* could not reproduce picture by Preston as zinc block or half tone but had engraver make copy. Engraver used imagination and added bend in river at left, as seen on opposite page.

by hiring the men from the other crew, and after breaking the jam too, got his logs down to Little Falls dam ahead of schedule.

In the lobby of the post office at Park Falls, Wisconsin, is a faded mural done by James Watrous, an artist employed by the Work Projects Administration in the 1930s. The mural depicts a wild fight between rival crews of drivers. Actually what the mural probably memorializes is the fight that probably took place when strikers confronted non-strikers and had nothing to do with who was going to get his logs down the river first as legend would have it.

As suggested earlier, many of the logs coming down the Chippewa River and its tributaries were absorbed by sawmills at Chippewa Falls and Eau Claire. Many more, probably twice as many, by 1880, were being driven all the way down the river into Beef Slough to be sorted and sent down to the coupling works of the slough below Nelson's Landing to be made up into brails. A brail was a huge raft of logs bound tightly together with cable and heavy wire which could be pushed down the Mississippi by steamboats to sawmills in Minnesota, Illinois, Iowa and Missouri.

The making of a brail was somewhat complicated. One eye-witness, Walter A. Blair, tells how it was done in his book, *A Raft Pilot's Log*, to wit:

> Each raft was composed of two pieces (halves) of three brails each. A brail of logs was six hundred feet long and forty-five feet wide. The rim was made of the longest logs, fastened at the ends with about a thirty-inch lap, by a short, heavy chain of three links. A two-inch hole was bored nine inches deep in each log, and a two-inch oak or ironwood pin, with a head on it was put through an end link of the chain, and driven hard into the hole in the boom log. These logs, so fastened, made a strong boom or frame (with just enough flexibility to suit the job) into which the loose logs were carried by the current, and skillfully placed endwise with the current, by men, using pike poles and peavies. Then one-half inch cross wires were placed and tightened, to hold the boom and logs together and prevent spreading.

Thousands of pins, more commonly called "plugs," were needed every year by the brail makers. The ones that were pounded into the logs went downriver and never came back. A small sawmill, located north of the iron bridge across

Buffalo River, manufactured these pins. It is still remembered around Alma as the "old plug mill."

Although there were log drives on the Chippewa through the Little Falls dam until August 1911, Chippewa Lumber & Boom Company probably made its last drive in the summer of 1909. It was customary for William Irvine, general superintendent, to drive his buckboard or buggy a short distance up the river to where the logs were being funneled into the main boom, and he often took his daughter, Ruth, (Mrs. Oscar Richter) with him. She remembers that the cooks arrived on the wanigan ahead of the drive. In personal conversation with the author at her home in Manitowoc, Wisconsin, she recalled that the cooks went ashore to dig a "bean hole" and started a fire to cover a big pot of beans which then simmered all day. When the crew came in from the drive in the evening, the men lined up for their plate of beans and fresh biscuits, and after they had eaten their fill, they stood around the fire and sang French-Canadian folk songs. "It was pure magic," Mrs. Richter sighed. "Pure magic."

William Hoyer told the author in personal conversation that wanigan (upper left, opposite) was last one used on Chippewa River on last drive for Chippewa Lumber & Boom in 1909. Under window are fresh biscuits. White aproned man is Mike McAleer, cookee, and other is Al Smith. Ed Gorgan, No. 1 cook, was not present. Hoyer described Gorgan as "best damned cook on *Chippeway* river." Note long oars used to guide wanigan, and at right, a batteau in tow. Belows, the *Dancing Anne*, another wanigan, at Eagle Rapids, 1909. Women are probably wives and friends of drivers and have come aboard, probably on Sunday, for lunch.

Opposite, upper left: wanigan running rapids. Lower left: breaking up jam on Big Eddy, above Chippewa Falls, 1906. Above, driving crew for Chippewa Lumber & Boom, William Hoyer stands second from right, pipe in mouth. At right of him may be Joe Violet, a Chippewa. Below, the *Sue Larson,* another wanigan for Chippewa Lumber & Boom. Left to right: George Kappus, Eli Robarge, cook, Walter Loiselle, Harry Fisk, Bert Glenn, Thad Loiselle, and Charles ("Kib") Ecker.

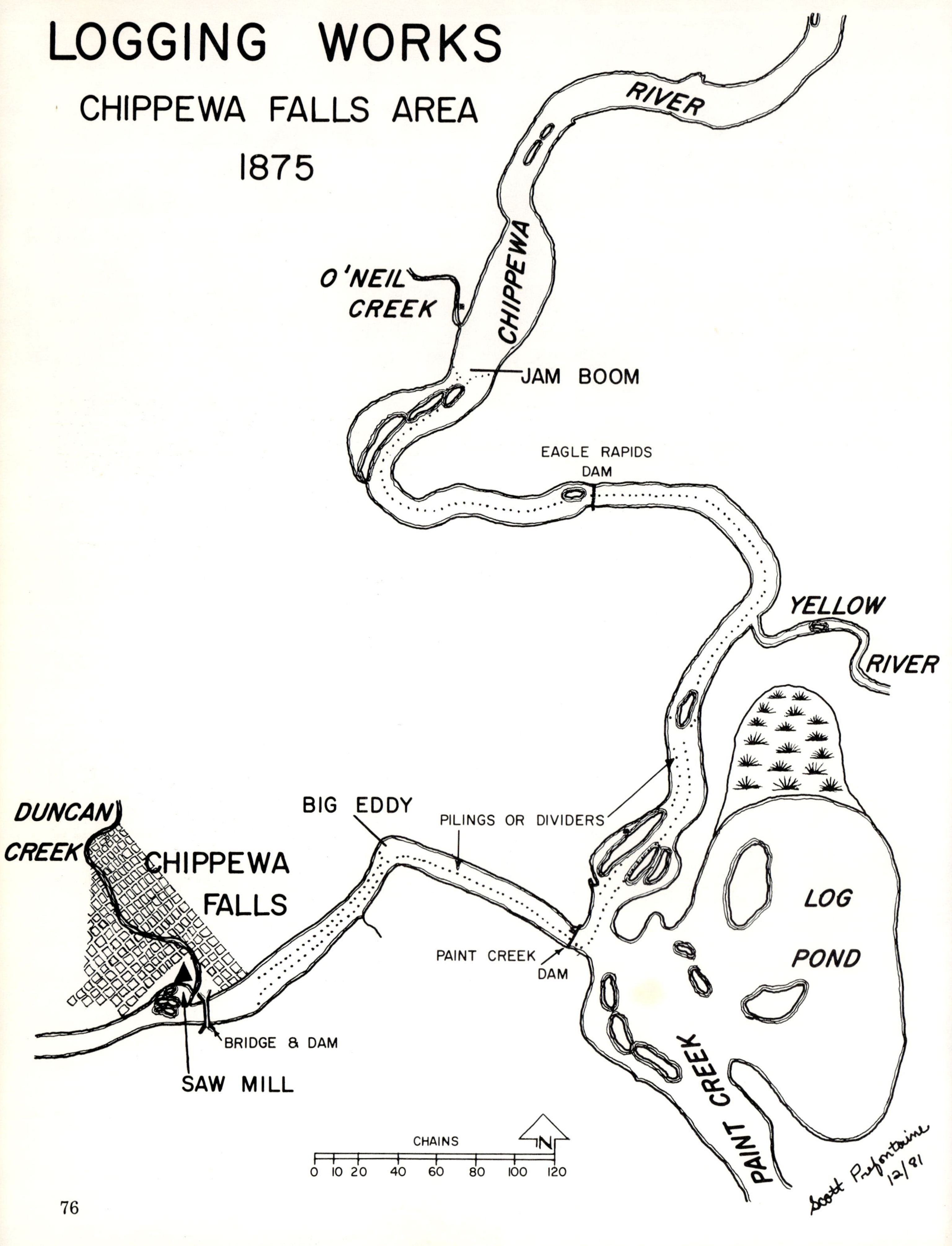
LOGGING WORKS
CHIPPEWA FALLS AREA
1875
RIVER
CHIPPEWA
O'NEIL
CREEK
JAM BOOM
EAGLE RAPIDS
DAM
YELLOW
RIVER
PILINGS OR DIVIDERS
BIG EDDY
DUNCAN
CREEK
CHIPPEWA
FALLS
LOG
POND
PAINT CREEK
DAM
BRIDGE & DAM
SAW MILL
PAINT CREEK
CHAINS
0 10 20 40 60 80 100 120
N
Scott Prefontaine
12/81

When this picture was taken there was little left of "Old Paint Creek Dam" built in 1870s. It stood in lower center where rapids rush through. Paint Creek entered Chippewa at right of picture. Upstream are two sets of jam piers. There were, no doubt, others which have washed away. Piers were used to slow up progress of logs and to create holding area for logs to be sorted as they came down from north. After sorting they were shunted into holding pond seen on map, and when logs were needed by mill owners, they were released. Area today covered by Lake Wissota.

Driving crew for Chippewa Lumber & Boom at Jim's Falls, 1909, ready to shove off. Below: driving crew in 1903, same company.

Cook from wanigan who is probably thinking, "They ate everything I had and left me with the dishes!" Below: abandoned camp for drivers on Thornapple River at Cameron Dam. Dam at left washed out in spring of 1906.

At one time Chippewa Lumber & Boom employed an entire driving crew made up of Chippewa Indians, but most of crews were white, mixed with Indians or French-Canadians of Indian descent. Above, three Indians holding pike poles to push logs. Note oar locks on batteau. Most of time men used poles to push boats. At left, a tepee in snow with two Indian families at Kennedy, near Park Falls, 1910. Man at left holds lever action rifle, probably Winchester. Wood for fire lies under canvas at right. Top of tepee is blackened by smoke from fire burning inside.

A double chain hoist for loading cars directly out of a river or pond. Steam engine in shed at left furnished power. Loader acted on same principle as bull-chain in sawmill. In left hand corner of picture runs boom of two square timbers with no planking. Picture was taken for Yellow River Lumber Company at Hughey, a landing on Omaha Railroad spur which ran east into Taylor County and terminated on Yellow River north of present Chequmegon Waters Flowage.

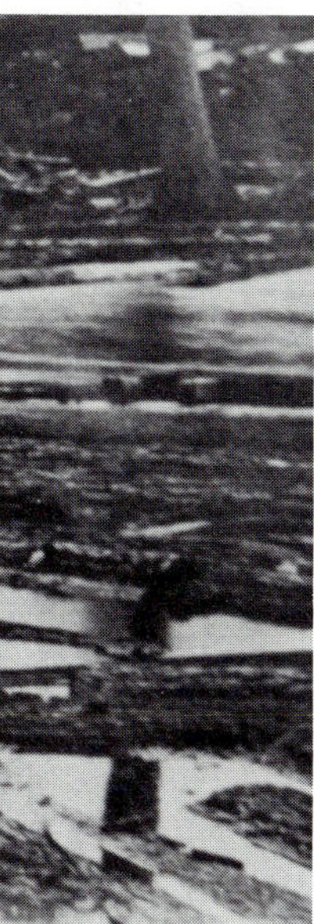

A sorting works on Yellow River, probably at Hughey. Makeshift bridge runs from shore to walking booms where two men in rear are at work. Man at left seems to be making notation, probably scaling.

Men and horses here are "dry rolling," that is, rolling logs stranded on dry land back to river. A special swivel, explained in advertisement below, was used on chain which was attached to single tree behind horse. With one horse on each end, log rolled like a wheel. In distance, village of Alma, and at right, a steamboat. Below, advertisement for dry roll kit in *Mississippi Valley Lumberman*, March 6, 1896.

Here's a Combination You'll Need in a Few Weeks.

It's For Bringing Runaway Logs Home.

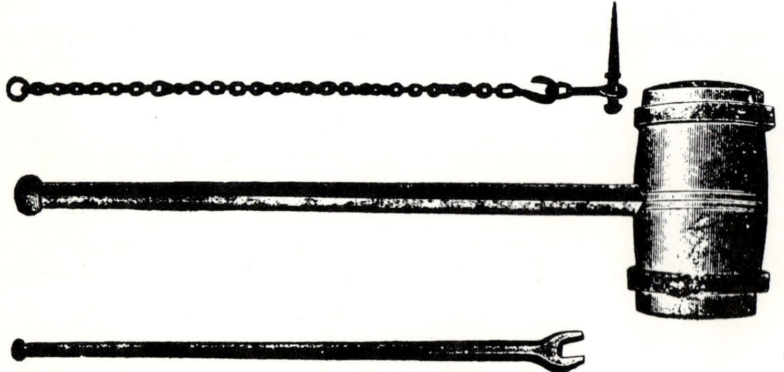

When logs are banked along streams with low banks, they sometimes in the spring, get carried away off over the low bottoms and there deposited when the water recedes. These three tools and a team will yank them all back to the river while you're GETTING READY to do it any other way.

It's a good deal cheaper way, too.

With the maul drive the pin, with chain attached, into end of log; put one in other end, too; hitch the team to chains, and you're off, the log rolling along like a wheel, the pins acting as axles. Arrived at the, river, take your claw bar and pull the pins—that's all. Go and do it again, and keep on doing it till your logs are all where you want them.

Easiest and chepest way you ever saw. The tools don't cost much, either.

Lumbering and Driving Tools

And Mill Supplies of all kinds. We have a splendid stock of everything required by lumbermen, and while we don't sell Cheap John goods at any price, our goods are cheap for the quality.

Write for anything you want, and our catalog.

EAU CLAIRE MILL SUPPLY CO., Eau Claire, Wis.

Above, fin sheer boom north of Alma on Mississippi River. This was reputed to be 1150 feet long. Picture was taken by Gerhard Gesell, studio photographer in Alma, probably 1889. Picture appears to be only one available of sheer boom in Wisconsin. One of crew sits on hinge of fin. Other fins can be seen farther along boom. Line ran from tip of each fin back to windlass and by manipulating line, boom boss could control position of boom. When line was tightened, fins swung out and forced boom into current. When boom had to be floated to shore to allow boat or raft through, line on fins was slackened and fins swung inward while current pushed boom to shore. Fin boom was invention of Levi Pond and James Allen of Eau Claire, 1861, and came to be widely used on rivers throughout United States and Canada. Picture above was probably taken at time when Weyerhaeuser moved sorting works out of Beef Slough over to West Newton on Minnesota side, and fin boom was probably from Round Hill.

Size of these pine logs is above average. They were cut in eastern Sawyer County and driven down Thornapple River to Cameron Dam to be sluiced and sent on to Chippewa Falls. But in spring of 1904, John F. Dietz, who owned part of dam, halted log drive and refused to permit Chippewa Lumber & Boom to drive. Logs floated in water for two years and in 1906, dam wash out and logs sank into mud. In picture above man in foreground with glasses is A. E. Roese, a newspaper editor from Osceola who came for deer hunting. Others, l. to r. are Leslie Dietz, Clarence Dietz, Harry Roese, and at right, Matt Egstad. Picture was given to author by Mrs. Egstad, 1960.

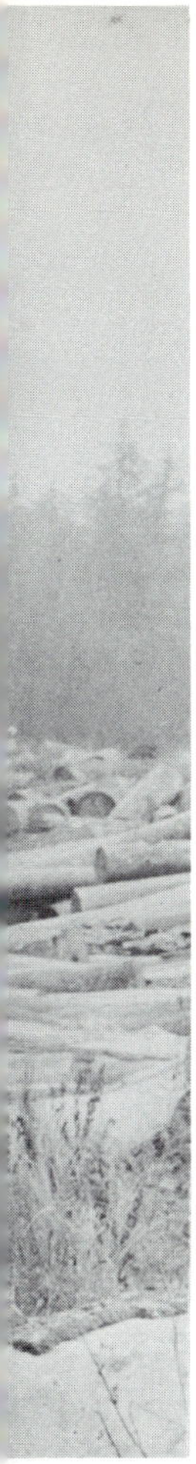

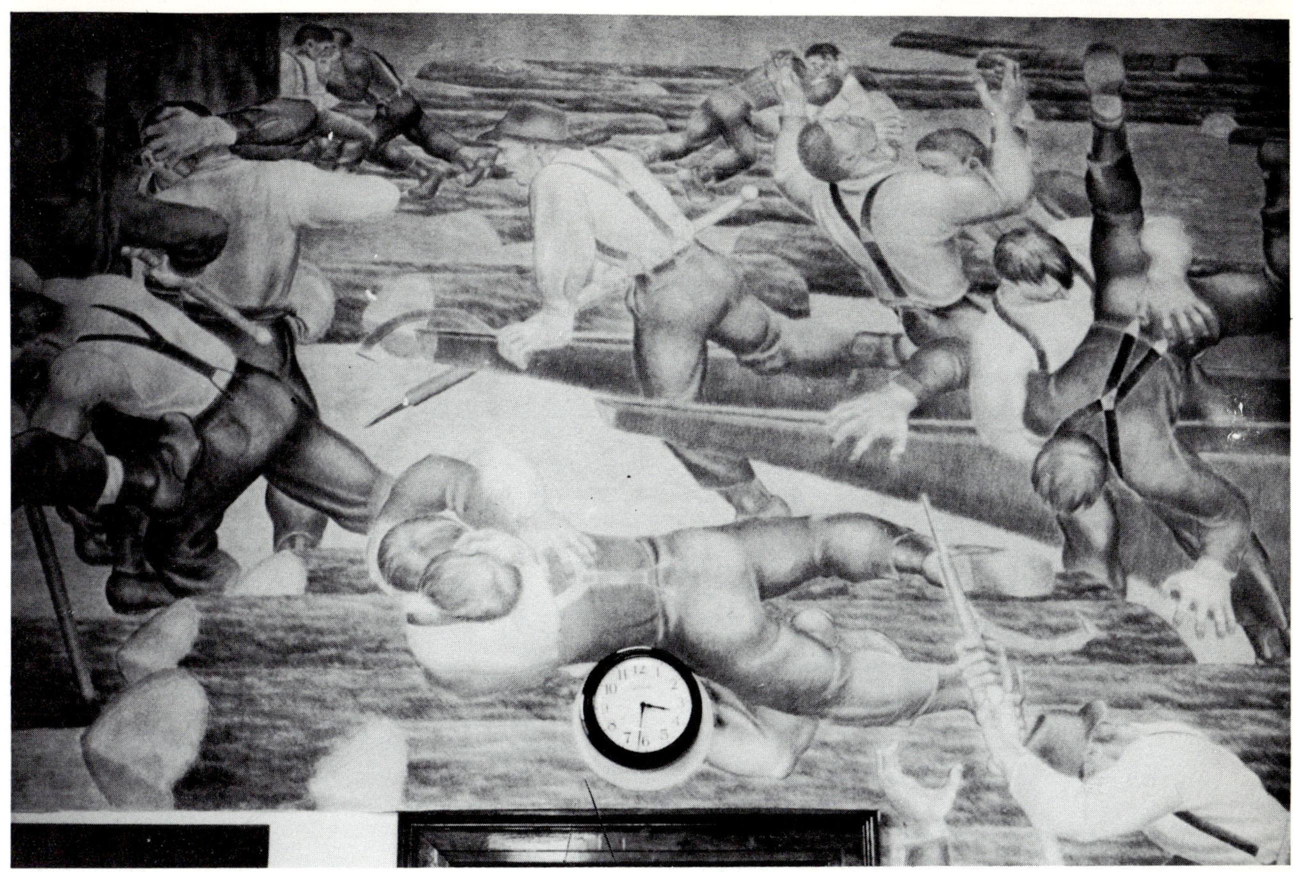

Mural by James Watrous memorialize fight between rival river crews near Pike Lake dam on south fork of Flambeau in 1888 or 1889. Mural may be seen in lobby of post office at Park Falls, Wisconsin. Below, last drive on Chippewa for Chippewa Lumber & Boom, summer of 1909. Archie Smith, drive boss, stands at right in white hat, avoiding camera. Man in derby may be William Irvine.

Above, sorting works and storage pond for logs above Dells Dam at Eau Claire in 1880s and 1890s. Ten million board feet of logs were processed through here every day in season. Jam piers held logs back to make room for sorting works seen at lower left. From here logs either were sluiced through Dells dam or funneled into canal running into Half Moon Lake. In center of picture is a pole with cross-arms fixed to jam pier. This same pole may be seen in picture on opposite page.

Gap on Dells pond where logs were sorted according to their ownership established by end marks and bark marks, and sent on to sawmills downstream. Pole with cross-arm at right may be seen in center of picture on opposite page. One of crew wields peavey (lower left), others pike poles to shove logs forward from walking booms seen at right and left. Plank bridge runs across gap in center. Crew probably found it safer than to "run" logs in loose pattern such as this.

Three members of a rear crew sacking a log into river. Man at left carries utensil to make tea. Tea leaves and sandwiches were carried in pouch barely visible on right hip. Source of picture uncertain.

Driving crew for Art Gore of Chippewa Falls and Charlie Stinson of Augusta, partners in contracting job on Clam River. Two men at left have fixed their peavies into same log to counterbalance their position. Photo was probably taken in 1902.

End Marks and Bark Marks

The term "end marks" used in logging lore refers to markings made at the ends of logs to identify their ownership. These markings were made on the logs before they were rolled into the river to be floated down to a sawmill, or on a flat car to be shipped to a mill on a railroad. The tool used for marking end marks was called a "stamp hammer." This had a three to eight pound iron head fitted on a wooden handle about three quarters of the length of a common ax handle. Into the cast iron head on one, or both faces, numerals, or small figural designs, such as a horse shoe, baby's foot or hat, were cast. The mark was cast backwards so that when the embossed design was struck into the soft wood at the end of the log, it left an imprint which could be read from left to right.

The scaler, who measured all the logs out in the woods, was usually responsible for striking the end marks into the log at least four times around a circle fairly close to the outer ring of bark. Smaller logs were usually struck once or twice in the center.

Howard Peddle holds end of log stamped with two end marks, one a company mark, ELX, other with two U.S. marks, latter denoting log has been cut on Indian reservation. In his right hand, Peddle holds typical stamp hammer to make end marks, this one called the "boxed J."

A stray log without an end mark was a "prize" with no legal owner. When unmarked logs reached a sorting works, they were shunted into a "bull pen" and later auctioned off to the highest bidder and the proceeds turned over to the boom company for storage. There were other ways but this was the most common practice for unmarked logs.

Before the log drives in Maine became big business, all that was needed to identify a log was a "bark mark," also known as a "water mark." An ax was used to cut through the bark and into the wood to make a combination of straight and slant lines. The swamper was usually responsible for these "hacks" as they were called. Relics of bark marks are rare, but several are preserved in slab pieces held by the Paul Bunyan Logging Camp at Carson Park in Eau Claire.

Meanwhile, by the 1840s, in the Maine woods, there were more logs coming down the river than a swamper could cut hacks into. A new way had to be found and out of this need emerged the system of symbols or abbreviations of letters from the alphabet to denote the name of the lumber company or log jobber. These symbols were forged in cast iron and mounted on a wooden handle to make a stamp hammer.

The bark marks, as suggested above, were made with a sharp ax.

Stamp hammers are believed to have been used in Maine as early as 1838, and in Michigan by 1842. When they came to Wisconsin has not been determined, but the marking system was definitely made a part of state statute in 1861, and again in 1864, when the state legislature enacted legislation to legalize title disputes. The 1861 statute also established lumber inspection districts on the Wisconsin, Black, Chippewa and St. Croix Rivers, referred to as districts 1, 2, 3, and 4. Each district had an inspector appointed by the Governor of Wisconsin for a two year term. (The first inspector of the Chippewa River district in 1861 was Hiram P. Graham.)

These statues authorized every log owner to use a certain mark, and further, it had to be recorded at the office of the district inspector for a fee of fifty cents. All unmarked logs were to be considered "prize" logs to be divided among the owners in each subdistrict. Anyone altering a mark was liable to a $10 fine. The act also made the registration of these marks a matter of public record, and required that the lumber inspector's scale bills should describe the mark on the logs measured, and any legal document in which logs were bought or sold would also use these marks. It was the duty of the lumber inspectors, further to:

> . . . . . record all mortgages, liens, and bills of sale, or other written instruments in any way affecting the ownership of any mark of logs in his district . . . *provided* that said instrument shall specify the marks placed upon said logs, and when they were cut, and shall be recorded in the office of the inspector in which the said marks are recorded . . . . and no conveyance, lien, mortgage or transfer shall be valid until the same is recorded.

The 1861 statute establishing the first lumber districts in the state was augmented by further legislature in 1863 establishing a Green Bay district, but exempting an area near Lake Winnebago and the Oshkosh mills. Finally, after years of confusion, the legislature, in 1878, brought the Oshkosh district within the system. An act of 1866, meanwhile, had declared the West Twin River and its tributaries in Brown and Manitowoc counties navigable streams and log owners were required to have their own end marks, but in this instance, the statute failed to order that the log marks be registered with the district inspector. And it was not before 1889 that legislation was passed requiring any lumberman on the Wisconsin River or its tributaries to have an end mark.

Nowhere does the language of the statutes use the term "end marks," a designation, apparently, of more recent origin. Nor is there any reference to the stamp hammer used to make these marks. Nevertheless, as one historian notes, these marks "were as effective in determining ownership as a real estate deed." End marks made with paint were also legal and a few were used.

By 1895, there were seventeen inspection districts, most of the new ones being carved out of the old, except two new districts covering the Wolf River Valley and the Menominee River valley in northeastern Wisconsin. No changes were made again until 1919 when the seventeen districts were abolished and replaced by four which divided the state into four quarters, the north and south line following the 4th principal meridian, and the east-west line following the third standard parallel north. Small loggers usually registered two or three end marks in a life time, but some of the larger companies had a hundred or more, each stamp representing a different jobber, or year of use. In the Chippewa River district, north of Eau Claire county, there were recorded, as early as 1872, about one thousand markings. Hundreds of these marks are preserved in the archives of the state historical societies of the Lake States. The largest collection of stamp hammers in Wisconsin is held by the Menominee Logging Museum at Keshena.

This system, which had been so important to the logging industry, was abolished in 1927 because, by then, big time logging was at an end.

Drawing of bark mark, also called water mark, and of markings used on ends of logs, cut by Chippewa Logging Company. Bark marks seen here were referred to in lexicon of river drivers as "girdle sextile girdle." Straight lines actually girdled half way around log. Sextile refers to heavenly bodies, but loggers used it to designate mark of six points. End marks, seen below, were registered with inspector of 6th Wisconsin Lumber District in October 1888, and in 1882 figured in law suit between Chippewa Logging Company and timber owner.

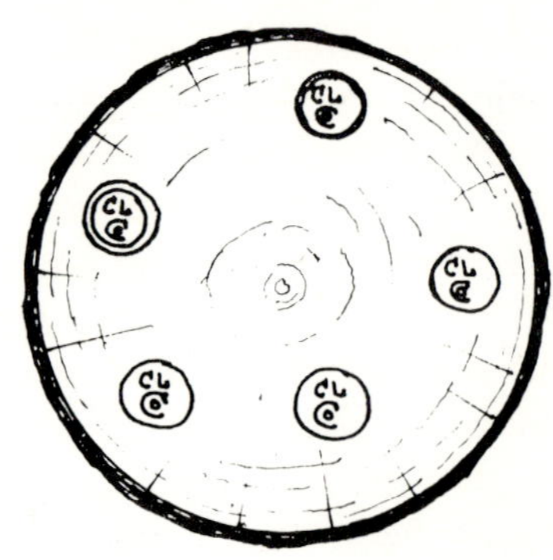

The log markings reproduced here, actual size, are found in a leather-bound note book kept by a man who signs himself as "Her Ole Evens," the "her" being the honorific for Mr. in Norwegian. He gives his address as "Buffalo, Wis." The name Evens is probably anglicized from Even, pronounced Aye-ven *in Norwegian. Whether immigrant or American-born, Evens had learned to be a log scaler, that is, he measured the number of board feet of lumber in a log. He worked on the rivers and landings all the way from Stillwater, Minnesota, to La Crosse, Wisconsin, and up the Chippewa to Eau Claire. It is not feasible to reproduce the entire note-book, but on the next three pages are some of the markings recorded by Evens, the bark marks in the left hand column, end marks on the right. Before he died, Evens probably gave the notebook to Albert Gilmore, boom boss at Round Hill, and it is now preserved by William H. Gilmore, a grandson.*

Robson

BARK AND	STAMP
AAP	AAP
AFT	AFT
ARN	Robson
CET	CET
C17	C17
CM	(CM)
DM	DM
DAK	DAK
GXL	GXL
HR	H R
HXW	HXW
HWL	H L
HXR	CE
MAK	MAK
MS	MS
MVJ (Eau Claire)	MVJ

Pioneer Lbr Co

[illegible]	[illegible]
[illegible]	PL CO
TX	[illegible]
HB	HB
X□	(X)
SW	□
S	(S)
F	F
MO	2X4
SOT	8OT
GKC	GKC
SGS	SGS
X□X (Eau Claire)	(B)

Northwestern Lumber Co

RMF	5/20
ANA	10/40
NY	NY
PNN	PNN
NXN	FFG
NAH	RF
8 (Eau Claire)	8

Rice Lake Lbr

RL	[illegible]
[illegible]	LR
KL	K7

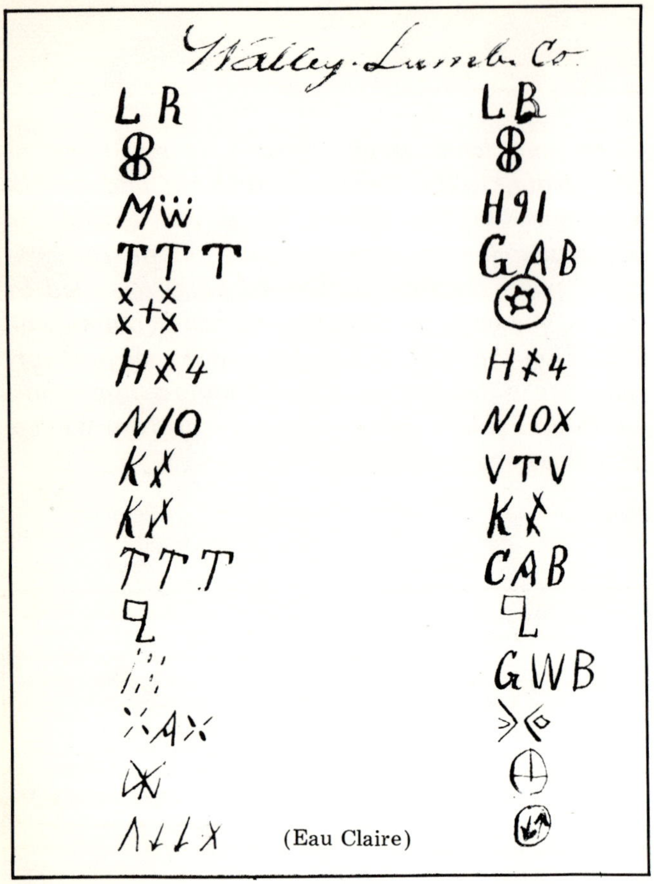

Valley Lumber Co.

LR	LB
8	8
MẅW	H91
TTT	GAB
x+x	
HX4	HX4
N10	N10X
KX	VTV
KX	KX
TTT	CAB
	GWB
XAX	
ALLX	

(Eau Claire)

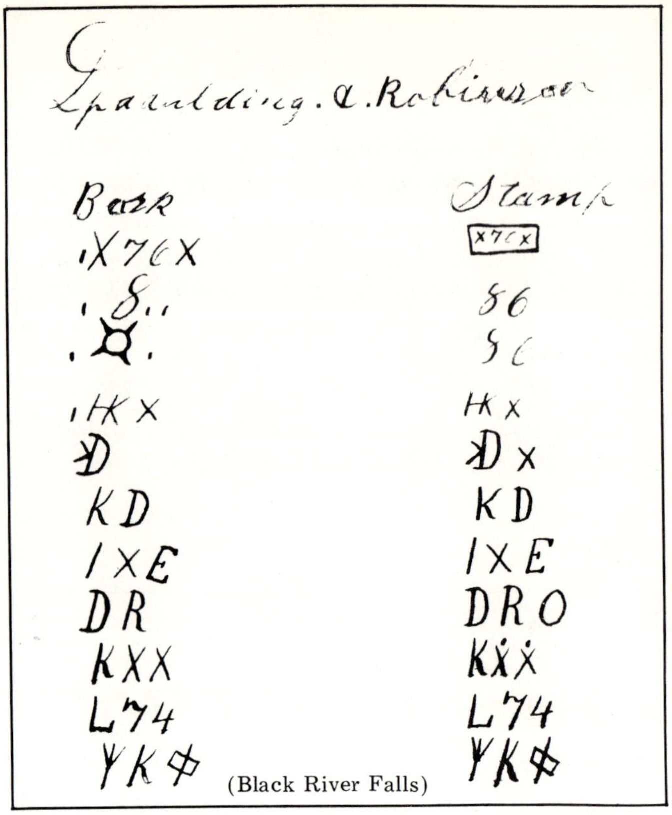

Spaulding & Robinson

Bark	Stamp
X76X	X76X
8	86
	96
HX	HX
D	DX
KD	KD
IXE	IXE
DR	DRO
KXX	KXX
L74	L74
YK	YK

(Black River Falls)

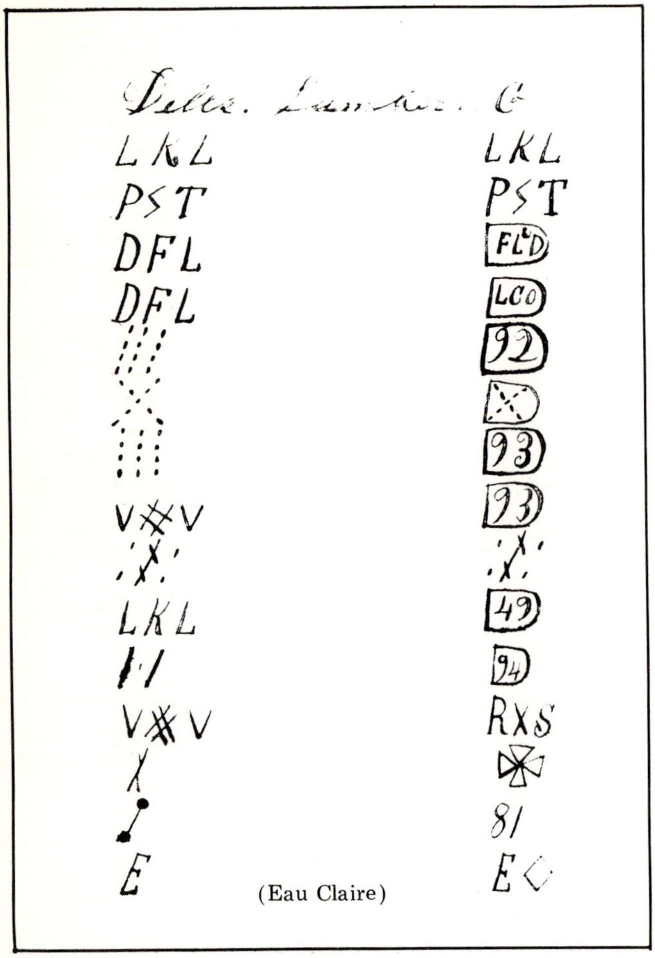

Dells Lumber Co.

LKL	LKL
PST	PST
DFL	FLD
DFL	LCO
	92
	93
VXV	93
X	X
LKL	49
II	94
VXV	RXS
X	
	81
E	E

(Eau Claire)

Mississippi River Log Co.

	E
S	TS
7A7X	7A7
KA*	CB
HH	HH
TIK	TIK
VT	
VHT	
FXW	FXW
BIT	BIT
YF	YF
AIM	XAX
A.IX	WJF
IR	W
F	F
KXE	KXE

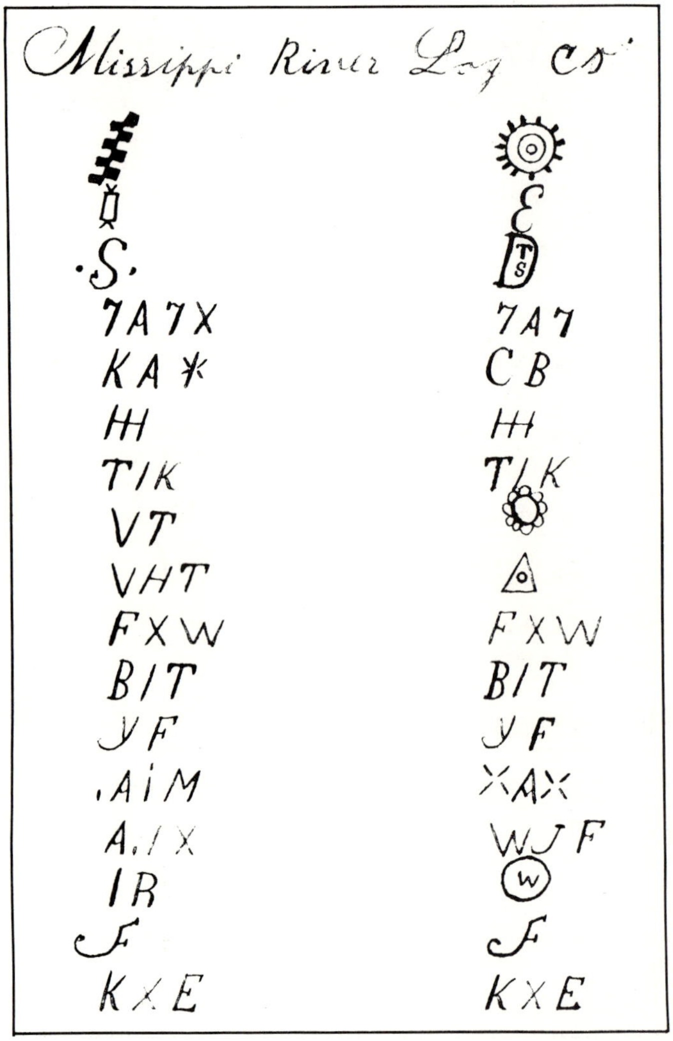

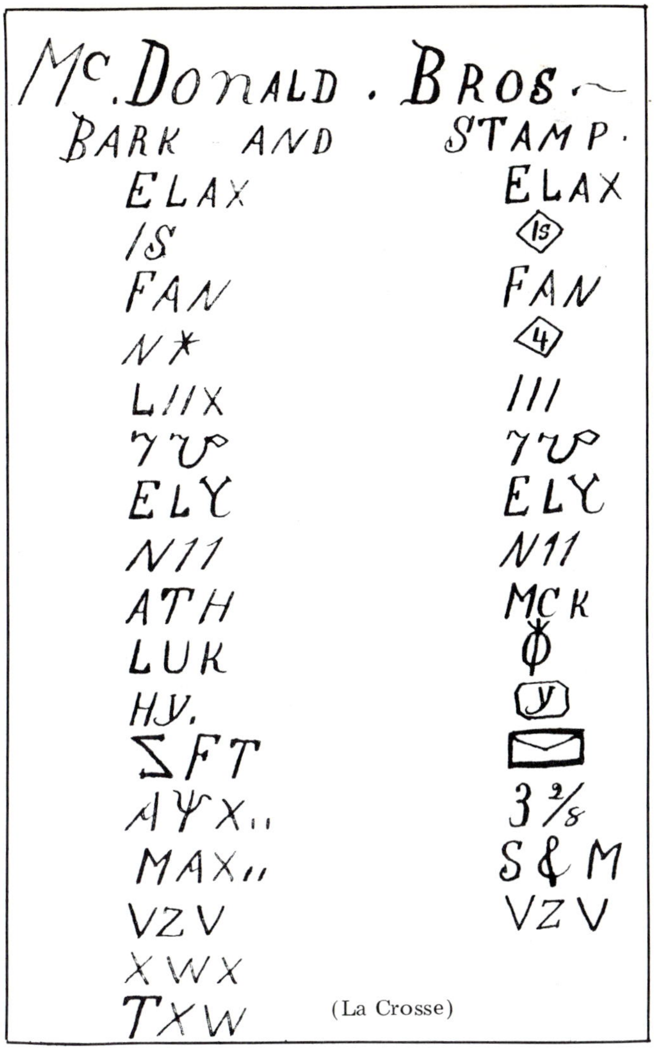

Mc.Donald . Bros.

BARK AND	STAMP.
ELAX	ELAX
IS	IS (in diamond)
FAN	FAN
N*	4 (in diamond)
L//X	///
7℧	7℧
ELY	ELY
N11	N11
ATH	MCK
LUK	ɸ
HY.	Y (in box)
ΣFT	(envelope)
AΨX..	3 2/8
MAX..	S & M
VZV	VZV
XWX	
TXW	

(La Crosse)

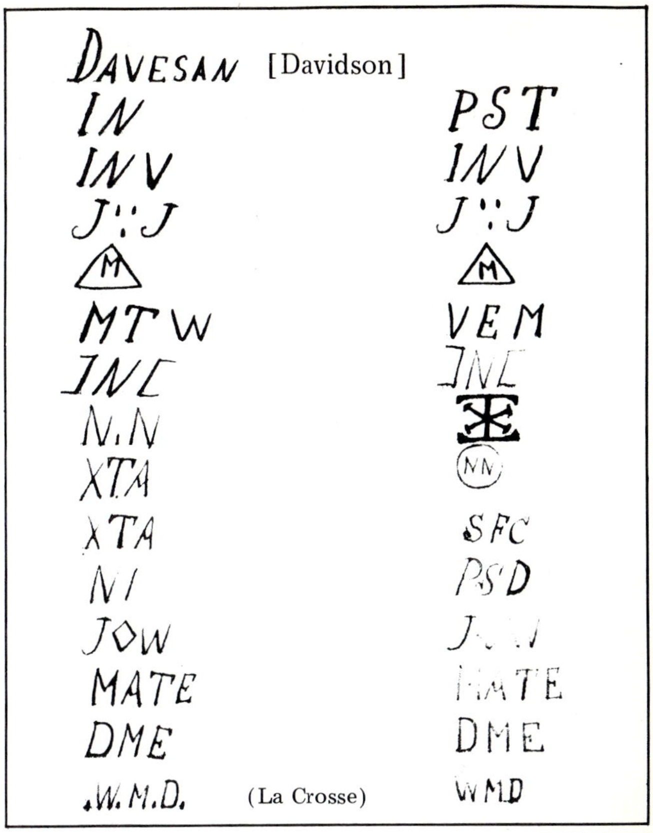

J. Paul

ΨWΨ	PAUL
WA	WAY
FAX	FAX
RUM	RUM (in box)
ȻE X	ȻE
ΛΛ	ΛΛ
ΨHΨ	PAUL
ΨKΨ	PAUL
ΨTΨ	PAUL
ΨPΨ	PAUL
ΨFΨ	PAUL
ΨEΨ	PAUL
Ψ///Ψ	PAUL

(La Crosse)

DAVESAN [Davidson]

IN	PST
INV	INV
J∵J	J∵J
M (in triangle)	M (in triangle)
MTW	VEM
JNC	JNC
N.N	(mark)
XTA	NN (in circle)
XTA	SFC
NI	PSD
JOW	JOW
MATE	MATE
DME	DME
.W.M.D.	WMD

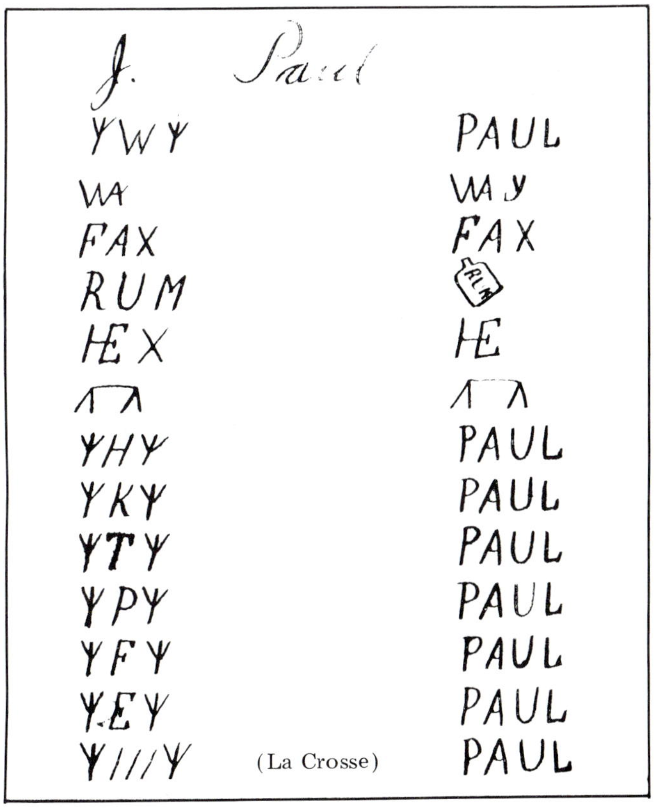

(La Crosse)

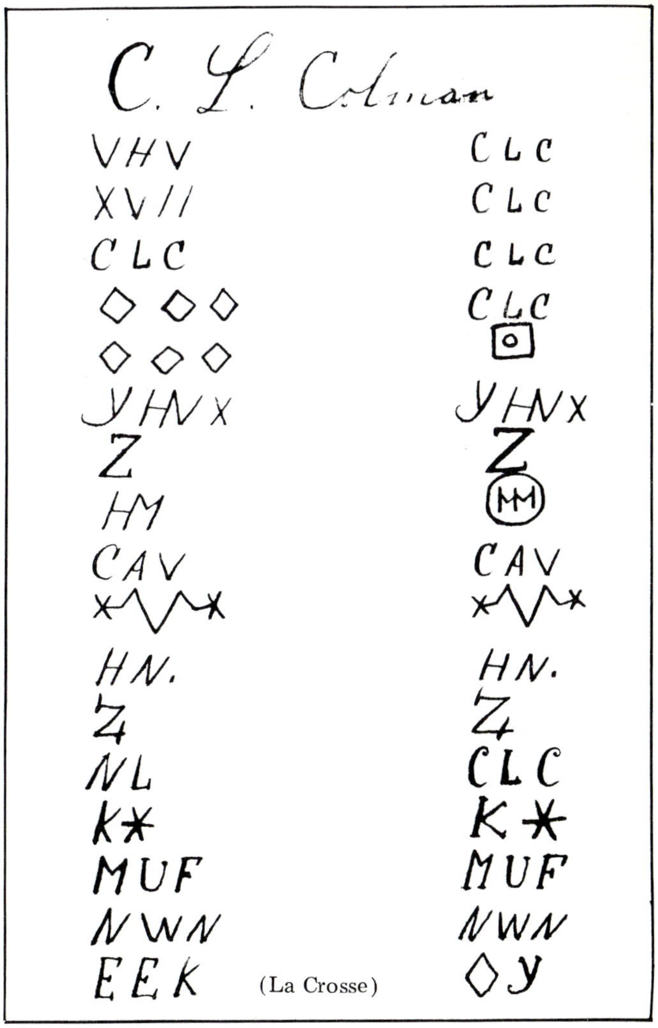

C. L. Colman

VHV	CLC
XV//	CLC
CLC	CLC
◇◇◇	CLC
◇◇◇	(square)
YHNX	YHNX
Z	Z
HM	HM (in circle)
CAV	CAV
X⋀⋁X	X⋀⋁X
HN.	HN.
Z4	Z4
NL	CLC
K*	K*
MUF	MUF
NWN	NWN
EEK	◇Y

(La Crosse)

Sawmill Operations

A century ago the city of Eau Claire was popularly known as "Sawdust City." In 1881, Eau Claire mills sawed 203,300,000 feet of lumber as compared to a total of 92,382,920 feet sawed in Wausau and Stevens Point. In the Lake States only Minneapolis surpassed the mill capacity of Eau Claire.

The early mill owners here and elsewhere in the Chippewa Valley began with "up-and-down" sawmills. On exhibit at the public museum in Downsville, south of Menomonie, Wisconsin stands the frame of an up-and-down saw. It was also called a "sash" mill because the frame which held the vertical saw blade resembled an enlarged window sash.

Down the hill and west from the museum in Downsville runs the Red Cedar River and it seems reasonable to assume that the relic in the museum was once part of a mill that was located on the river nearby. The sawmill operated on waterpower, either by overshot or undershot wheels, fed by water in a flume along the shore of the dam site.

This type of sawmill, variations of which were used in Europe for several hundred years, and in the American Colonies, could saw at least 2,000 feet of lumber in twelve hours, and if everything was running smoothly, it was possible to saw up to 5,000 feet in twenty-four hours. The output was small not only because the saw moved less than five centimeters per stroke, but because the log on the carriage had to be squared off first with a broad ax, at least on two sides, to give the primitive "dogs" a better grip on the log. A crew of two or three men was all that was required to operate the mill, and it was apparently not difficult to find a crew for the night shift. The mill ran in a leisurely manner; it could not be speeded up, and the crew in charge often played cards to pass the time of day or night.

Thus any discussion of saw milling in the Chippewa Valley before 1850, when the first circular saws appeared on the market, is bound to be associated with the up-and-down sawmills. As with most innovations, old techniques do not die easily even if the new are manifestly better. It takes time and money to change, and it often takes courage on the part of the owner to make the

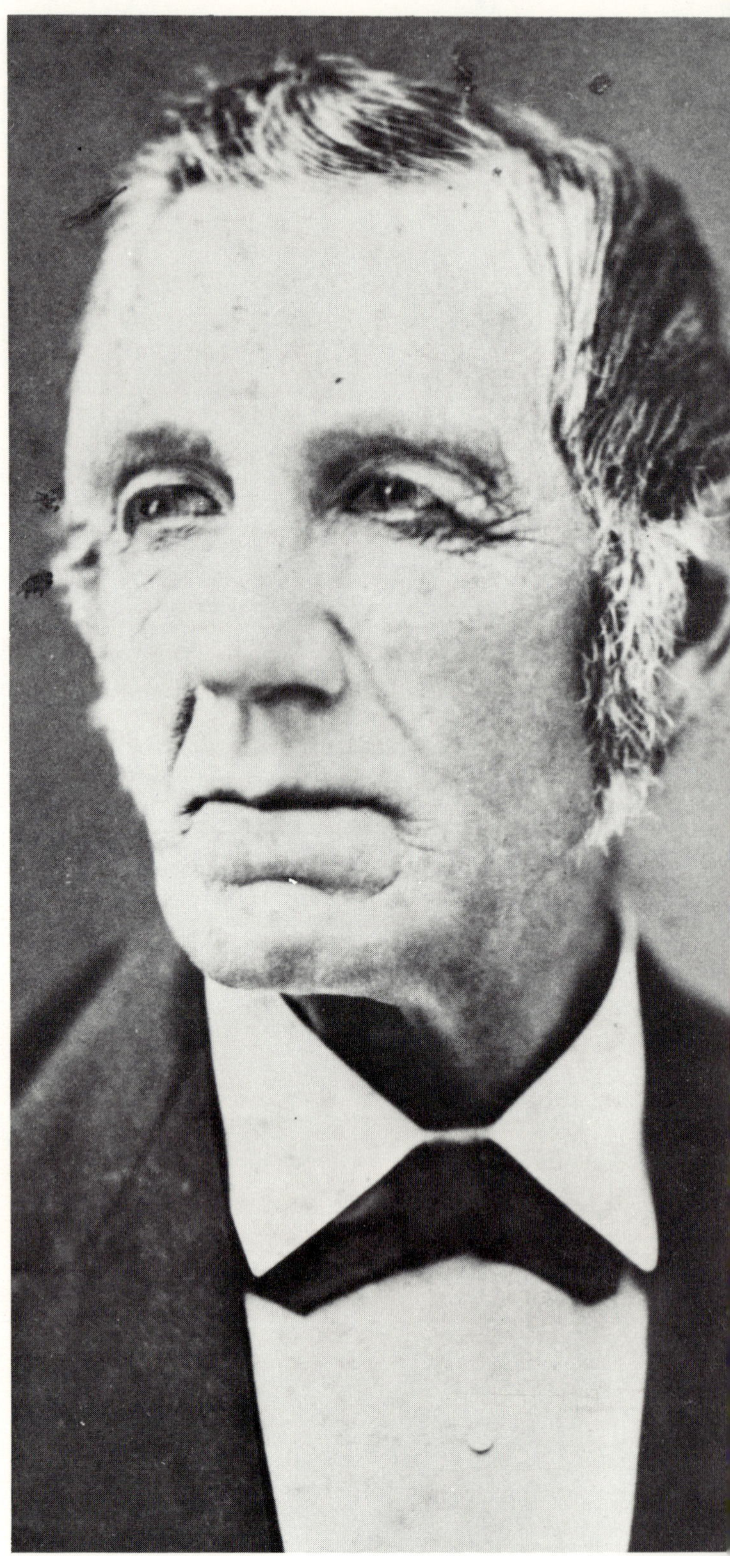

Hiram S. Allen, pioneer sawmill operator and one of founders of city of Chippewa Falls.

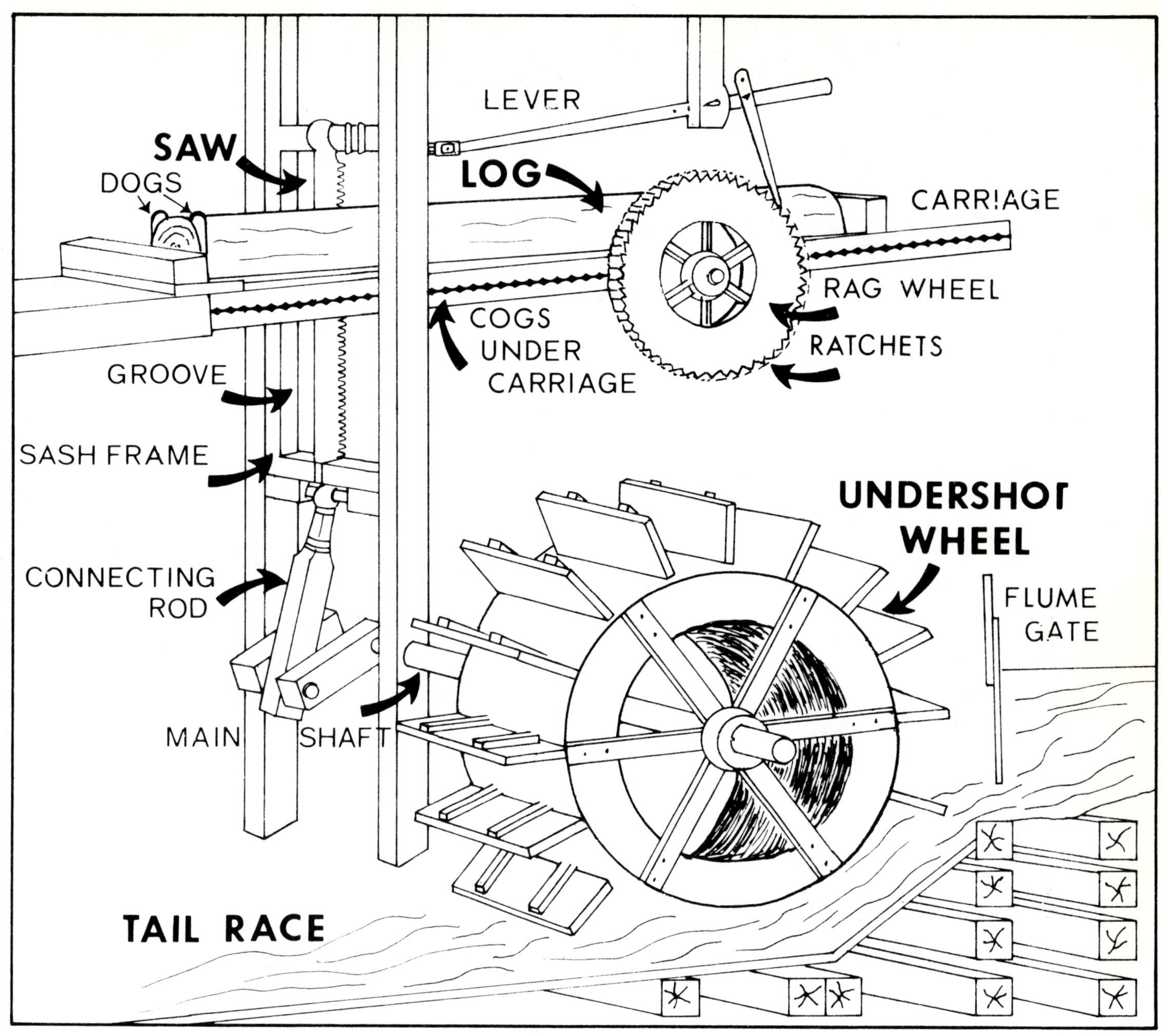

Drawing of an "up-and-down" sawmill by artist Randall E. Rohe. This was type of mill H. S. Allen first used in Chippewa Falls. Log was sawed by vertical saw within frame and entire frame moved up and down on connecting rod geared to shaft on waterwheel.

change. For this reason, it seems possible that the frame on display at Downsville could have been used on the Red Cedar as early as the 1830s but not later than 1870 when most mills had converted to circular saws and gang saws powered either by water turbines or steam engines.

The up-and-down saw was aptly named because the saw was, in a sense, a crosscut saw suspended in a frame. It was often called a reciprocating saw. It moved up and down *with the frame* and sawed only on the downstroke because the teeth were angled in one direction—down.

For the beginnings of the sawmill industry in the Chippewa Valley reliance has to be placed on recollections of pioneer lumbermen, since documentary materials are scarce. Perhaps the first account of very early sawmill operations was written by James H. Lockwood for the second volume of *Wisconsin Historical Collections* published in 1856. Lockwood was recording what he remembered from thirty years earlier and, while memories are unreliable, this account has to be given considerable credence. Lockwood was trying to remember, in essence, who was the first to build a sawmill in the Chippewa Valley and this is the way he describes it:

Interior of an "up-and-down" sawmill. Reciprocating saw blade is barely visible in center of sash frame. Wheel at right probably activated carriage seen in rear. Source of this picture uncertain.

> About the year 1822, a man by the name of Hardin Perkins, from Kentucky, came to Prairie du Chien for the purpose of building a sawmill in the Indian country, and obtained permission from Major Talliaferro, then agent for the Sioux Indians, with the consent of the Indians, to erect a sawmill on their land in the Chippewa River and tributaries; but Perkins not having the capital to carry out this project, or sufficient influence to obtain permission of the Indians to erect this mill, solicited Joseph Rolette and myself to join him, which we did, and contracted with [Chief] Wabashaw's band of Sioux, who claimed the Chippewa River country, for the privilege of erecting a mill and cutting timber for it, paying them about $1,000 per year in goods . . . He proceeded to Menominee [Red Cedar] River . . . and on a small stream running into the Menominee, almost twenty-five miles from its mouth, erected a sawmill and had it so near done that he expected to commence sawing in a very few days, when one of those sudden freshets . . . swept away the dam, mill and appendages, and Perkins returned to Prairie du Chien with his family and hands . . .

If the Red Cedar were measured by its actual course, bends and curves, nearly twenty-five miles would put the mill at what is today the city of Menomonie, and other accounts agree that the mill actually stood at the mouth of Wilson Creek. Lockwood and Rolette were having difficulties, however, with Colonel Josiah Snelling, commanding the fort of the same name, about their authority to build on Sioux land, but after clearing the matter of trespass with the Secretary of War, they were given permission once again to erect a sawmill. This was in the spring of 1830, and in May, they sent a man called Armstrong, a millwright, and also a blacksmith, carpenter, and several workmen back to the Red Cedar. Armstrong selected the same site as Perkins on Wilson Creek.

Apparently the mill was largely completed when hostility, or assumed hostility from the Sioux, frightened Armstrong and he returned to Prairie du Chien much disturbed by what he thought had been a threat on his life. Lockwood wasn't so sure. He decided to take Armstrong and a Winnebago mixed-blood and canoe up the Mississippi and Chippewa rivers into the Red Cedar to see what he could do about the matter. Armstrong caused so

Eau Claire Lumber Company sawmill on north bank of Eau Claire River was located about 100 meters east of Dewey Street bridge. Building at immediate left is rafting shed where lumber cribs were assembled. Cribs slid down ways, and two cribs apparently have just hit water. Smoke stack looms farther up river, site of second dam and water powered shingle mill for same company. Third stack represents planing mill. At right, flour mill. Below, photo of same sawmill taken in 1873, shows mill had already been converted to steam. At right, probably original flour mill on this site. Compare above. Both probably operated off water turbines.

1. County Court House.
2. Seminary.
3. Public Schools.
4. City Hall.
5. Engine Houses.
6. R. R. Depot.
7. Post Office.

CHURCHES

8. Methodist.
9. Congregational.
10. Baptist.
12. Methodist.
13. Episcopal.
14. Scandinavian Lutheran.
15. " "
16. " "
17. German "
18. Roman Catholic, (German).
19. " " (Irish).

EAU C

Birdseye view engraving of Eau Claire dated 1880 from photograph taken before 1878, since Dells dam does not show. Smoke from Eddy Mill may be seen in upper left. Steamboats were probably on their last run, but lumber raft is passing under Grand Avenue bridge. There are logs in Half Moon Lake at left but no sawmills, since picture is incomplete. Inset in lower right is key to West Side mills in 1880, or earlier.

much trouble en route that Lockwood agreed to buy his third interest, and Armstrong returned south, presumably in another canoe. Isaac Saunders, one of the carpenters who helped build the first mill, was taken in as a new partner, and Saunders hired another man named Holmes to complete the mill left unfinished by Armstrong. By March, 1831, writes Lockwood, this up-and-down sawmill had sawed 100,000 feet of lumber products, suggesting the mill sawed about 2,000 feet in a 12-hour day, with no night shift working.

Lockwood now ran into another problem: how to transport the sawed lumber down to markets on the Mississippi River. The Canadians he had working for him had only a vague idea how to make a crib of lumber, but a Yankee from New England remembered something he had seen in Maine, and together with Lockwood's own ingenuity, they managed to assemble a number of cribs to make a small raft. The problem was that the river level was too low to float a raft. And then, instead of a normal rainfall the river was suddenly deluged by a heavy rain. This drove Lockwood's rafts against rocks and shore lines, scattering lumber all over the bottoms of the lower Chippewa River. Less than fifty percent was salvaged. Thus, the first attempt to bring pine lumber from the Chippewa Valley into the Mississippi was a near disaster.

While waiting for the river to rise, Lockwood had put his crew to work on a second sawmill located off the Red Cedar, this one on Gilbert's Creek, a short distance below the mouth of Wilson Creek. This mill was probably operating later that year.

The Lockwood-Rolette-Saunders partnership was dissolved in 1835 with the sale of both mills on Wilson and Gilbert creeks to Hiram S. Allen, an experienced lumberman from Vermont who was shortly joined in partnership by George S. Branham.

On July 29, 1837, Chief Hole-in-the-Day and forty-seven other chiefs of the Sioux nation, signed a treaty at Fort Snelling to give up most of their lands in Wisconsin while retaining the privilege of hunting and fishing and gathering wild rice. Between 1837 and 1842, the Chippewa nation also gave up most of its lands in Wisconsin. Now there would be no more problems about the white men trespassing on Indian land; he could buy this land and the timber standing upon it from the government. One of the first partnerships to take advantage of the new dispensation was that of Jean Brunet, Hercules Dousman, Henry Sibley, William Aitken, and Lyman Warren, all former agents of the American Fur Company. Warren was also a former Lake Superior trader based at La Pointe, and he was probably taken into the partnership because of his knowledge of the pine country and because he was part Indian.

Jean Brunet who operated a tavern-inn at Prairie du Chien in the 1820s, was a Frenchman who had come up the Mississippi River via New Orleans. He seems to have moved into the Chippewa country in the late 1820s to trade with the Indians at a post on the west bank of the Chippewa River later identified with his name, Brunet Falls. Because of his wide acquaintance in Prairie du Chien, and the Chippewa country, the partners Dousman, Sibley, Aitken and Warren agreed to delegate Brunet to superintend the construction of an up-and-down sawmill at Chippewa Falls.

Although building the framework and carriage for an up-and-down sawmill required more muscle than gear, most early builders also needed some knowledge of hydrodynamics, that is, how to pick a good site on a river and how to utilize it to the best advantage. A builder also had to know something about booms to store the logs which would be coming down the river from the forests "up north." Scarcely any one of these early builders had any idea of how high or how fast the river they were trying to harness could rise.

Aside from problems associated with the river, the builders were plagued by a dearth of even semi-skilled workmen who could handle the problems mentioned, for any man worth his salt could command good wages in the booming frontier towns along the Mississippi. Hence, the work force available in the wilderness frontier of Wisconsin consisted mostly of drifters, army deserters, unskilled laborers, and old voyagers from Canada who loved the trees, but did not like building mills to saw them up.

Above, sawmill at left is probably Sherman Lumber Company which was built on north shore of Half Moon Lake about 1880. Mill operated on steam, and close-up lense reveals bull-slide with two logs on it, but probably no bull-chain. John Paul of La Crosse probably invented bull-chain in 1877. Chain operated on same principle as an escalator. Logs were snubbed against chain in the "hot pond" and conveyed to second deck of mill where sawing was done.

This photo was taken same time as one above. It shows mill of Eau Claire Manufacturing Company on east shore of Half Moon Lake. Mill is shut down, or it may be Sunday, since no steam comes from stack. Small white building at extreme left appears in top photo and it appears to have been painted in meantime. Logs stored here were not only for above mills, but for mills along Menomonie Street.

The site chosen by the Prairie du Chien partners at Chippewa Falls was unfortunate. Either Jean Brunet or someone else was so impressed by the possibility of the great waterpower available that they refused to see the mechanical difficulties of locating here. As a result, the mill was a money loser for many years before it turned a profit.

Work on the mill did not actually begin until 1838, and it was not before 1840 that any lumber was produced. A millrace was built 120 rods long, seventy-five rods of which ran through solid rock. Work on the mill race alone probably set back the time table for completion by more than a year. However, when finished, this race, or flume, created a twenty-two foot head of water without building a dam. The mill machinery was operated off a shaft turned by a flutter wheel, either overshot or undershot. This was probably replaced in the 1860s when water turbines became practical.

Parts for the mill and provisions for the crew had to come from Galena, Illinois, or Prairie du Chien, and breakdowns at the mill usually meant a trip to either city for parts. Jean Brunet, who had taken on the job of construction, got blamed for mistakes and delays. Dousman took the lead in dissolving the partnership when he sold out to Jacob Bass and Benjamin W. Bronson, both of Prairie du Chien, for $11,000, at which time Dousman wrote Sibley and said: " . . . . thank God it is off our hands and I consider myself well rid of the infernal concern even if I never get a cent more—thinking of it has made me nervous so no more at present." Dousman would build Villa Louis on peltry, not the pinery!

However, since two strangers were willing to risk $11,000—a small fortune at the time—on an ailing sawmill suggests that it was by now a rather large operation. It had been converted from the up-and-down saw to a muley saw operation. The muley resembled the up-and-down saw in that the saw blade was suspended vertically over the log, but the difference was that the mill shaft lifted the muley saw blade up and down *within* the sash frame, not the entire frame. The muley also wasted less lumber because it sawed a smaller kerf. There were two muley saws in the mill which were capable of sawing more than 15,000 board feet a

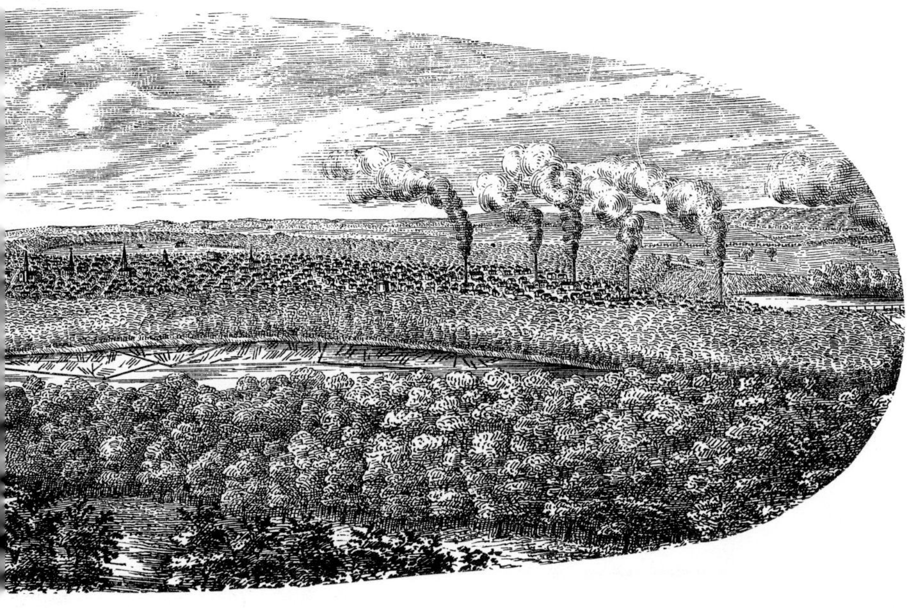

Birdseye view of Eau Claire from an engraving appearing in *History of Northern Wisconsin* (1881). Scene looks east across Half Moon Lake, but in order to include terrain and mills in distance, engraver has "lowered" Carson Park hill. Lake has logs in it and at left are mills of Eau Claire Lumber Company on Eau Claire River and probably Pioneers Lumber Company mill at far left. At right, mills on Menomonie Street. Engraving stresses smoke coming out of stacks to show readers what a prosperous city this is.

The Northwestern Lumber Company mill on the West Side. The vehicle seen here was called a "dump truck," or "dump cart," and it was used mainly for hauling sawdust, or shingle wood to homes around city. Cart has gear to tip cart backwards. One team of horses has housers over hames, the other not.

John H. Kaiser Lumber Company mill on Tenth Avenue next to the 1857 cut-off. Here bull-slide has chain to convey logs to second deck of mill where sawing was done. In foreground two pond monkeys shove logs towards bull-slide. Stack has spark arrester. Kaiser logged mostly in Sawyer County in the early 1900s and cut mostly hardwood.

Lee Lightfoot in engine room of Kaiser Lumber Company.

day, but source materials fail to explain whether this was for a 12-hour or 24-hour day, most likely the latter.

When Bronson and Bass took over the mill, the machinery in it was run down. Also, there were no piers, booms, or a pond to hold incoming logs, and the race needed repairs. The new owners apparently bought on faith, not on good sense, but, instead of a loss, the venture was a success, mainly because they had been joined in partnership by H.S. Allen and Branham, both experienced lumbermen who had been operating mills over on the Red Cedar. Allen, not satisfied with the two mills on Gilbert and Wilson creeks, built a third mill on Spring Creek, and finally, in 1839, a fourth mill on the west bank of the Red Cedar three kilometers or so below Gilbert's Creek. It was at this time that Allen took in George Branham as a partner. In 1841, they sold the Spring Creek mill, and a fire destroyed the mill on the Red Cedar in 1843. In 1845, the two partners sold the Wilson Creek mill, and in 1846, sold the last of their Red Cedar properties on Gilbert's Creek to Samuel Gilbert. (The creek patently took its name after, not before Gilbert bought the mill.)

With the capital acquired from the sale of these several deals, Allen & Branham moved over to the Chippewa River to look for new opportunities, and settled on the Lower Dells in the present city of Eau Claire where there was a natural holding pond flanked by cliffs and high banks. Here they built a new mill and took into the firm two new partners, Simon and George Randall. Future plans called for cutting a canal through the west bank into Half Moon Lake for additional log storage.

However, another version of this "first" appeared in an article which ran in the Eau Claire *Free Press* on July 11, 1867, recounting the early history of the city. Here is the way the newspaper editor explained it:

> "The honor of the first settlement of Eau Claire 12 miles below the Falls [Chippewa Falls] is claimed by one Stephen McCann, a big, stout, good-natured Kentucky Irishman. His cabin was put up in 1846. His wife was the first white woman to remain in the Chippewa valley while his son Stephen Jr. was the first one sacredly 'immersed' in the Chippewa waters . . . He [McCann] helped put up the 'Blue Mill' six miles above Eau Claire in 1842. Simon and George Randall got out logs for the mill the next winter. Thomas Randall, another brother, came up to Eau Claire in 1845. He chartered a boat in Galena for $700 to take up his goods as far as Wabasha. At that time Nathan Myrick and V.J.B. Miller had an Indian trading shanty at La Crosse . . . .

From this it is not clear who actually built the first mill above the Dells. McCann "helped" put up the Blue Mill, but other sources believe Simon and George Randall and J. O. Thomas were the others involved in this. The mill was carried away, however, by flood in June 1847 and the Randalls allegedly rebuilt it.

A second sawmill was erected in the spring of 1847 by Reed & Gage, according to the account quoted above, but no explanation follows, and no other source ever mentions Reed & Gage in this connection. Also, the article quoted above makes no mention of H.S. Allen or Branham in connection with any mill built on the Chippewa either at the big eddy or below the Dells where the city of Eau Claire developed. There seems to be agreement, however, that Allen left his partners in Eau Claire to join his fortunes with Bass & Bronson at Chippewa Falls. Here he immediately began making improvements to the one-time mill of Jean Brunet.

The spring of 1847 was dry and logs were lying on banking grounds up the river unable to be moved. Then, as so often happened, the flood came. It began to rain on June 15th, and by the following morning, the river had risen twelve feet. The flood tide picked up thousands of logs and carried them downstream, swept through Chippewa Falls in one giant avalanche taking everything with it, piers, booms and sawmill.

In 1848, in an attempt to get more capital, Allen organized the Chippewa Falls Lumber Company and began rebuilding the mill, this one larger than the first, and by 1853, it had five muley saws and two circular saws. By 1855, it was assumed that the company had turned the corner and left misfortune behind. Then disaster struck again. On July 6th, a storm raged over the valley for more than a day and the rise in the river brought down thousands of logs which swirled past Chippewa

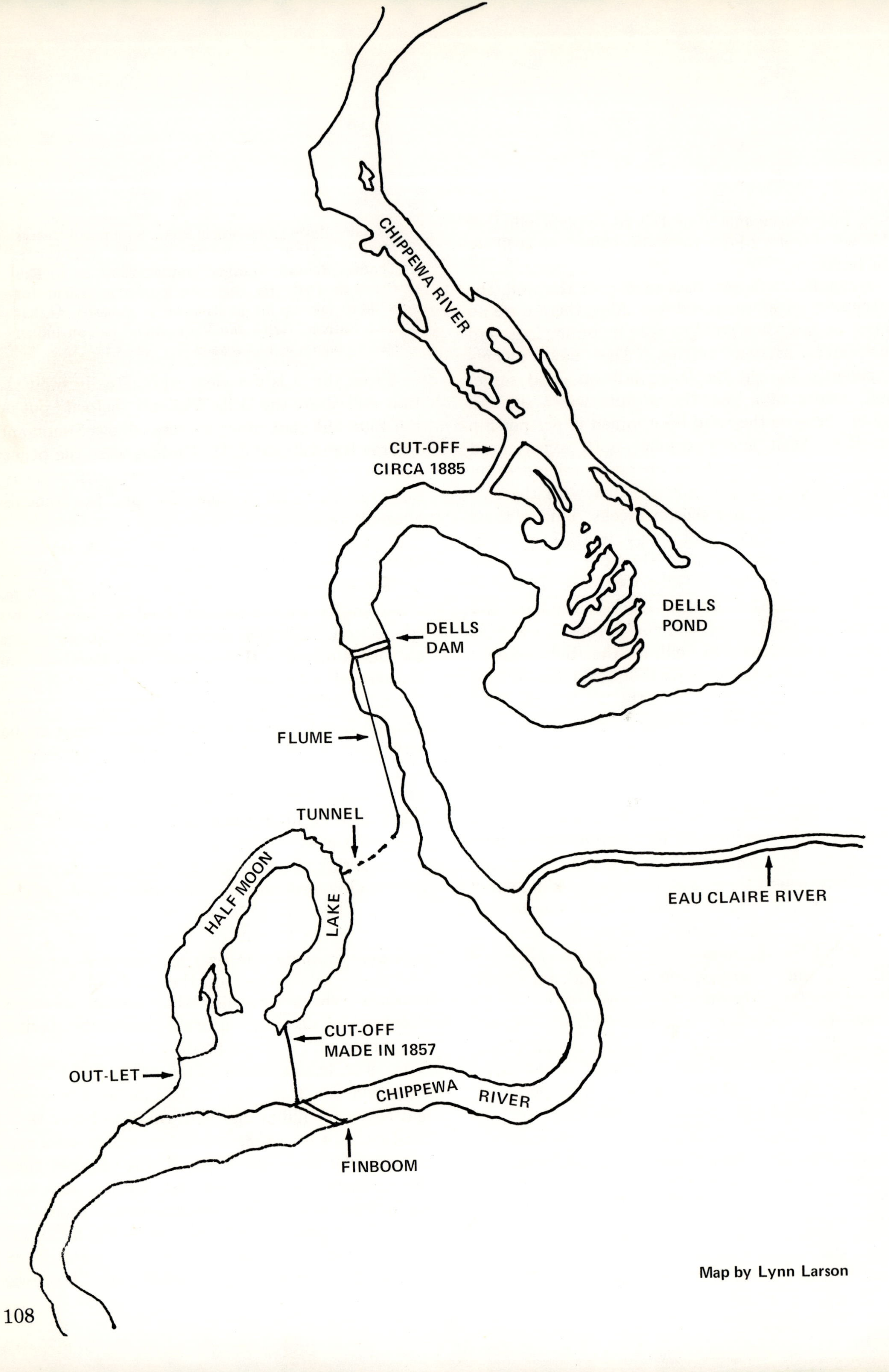
CHIPPEWA RIVER
CUT-OFF
CIRCA 1885
DELLS
POND
DELLS
DAM
FLUME
TUNNEL
HALF MOON
LAKE
EAU CLAIRE RIVER
CUT-OFF
MADE IN 1857
OUT-LET
CHIPPEWA
RIVER
FINBOOM
Map by Lynn Larson

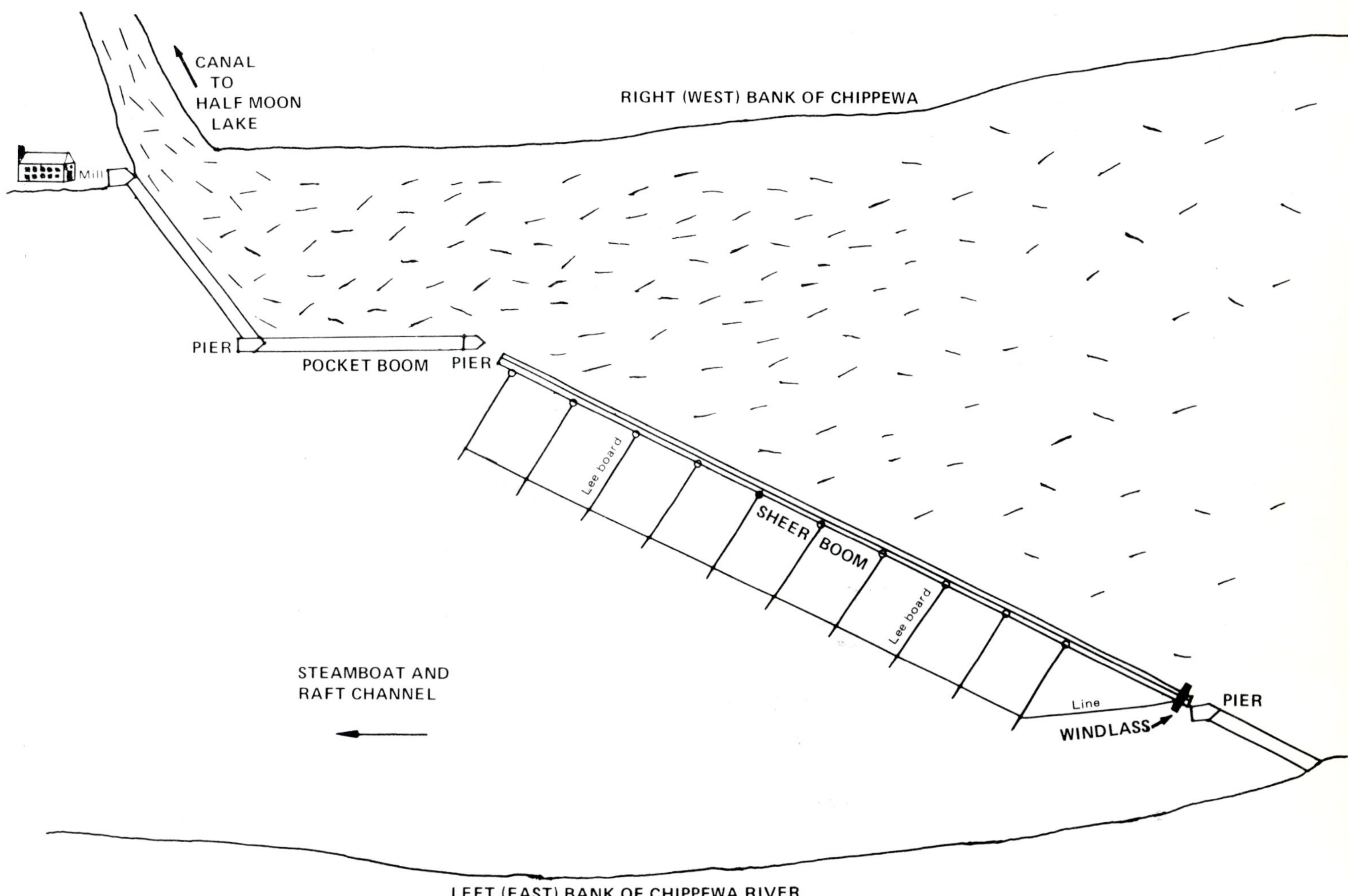

Opposite left: map of Chippewa River showing 1857 cut-off to Half Moon Lake. Also flume to Half Moon Lake built in 1879-80, and cut-off at big bend of Chippewa above Dells pond. Above: sketch of fin sheer boom which shunted logs into 1857 cut-off. Fins are extended perpendicular to boom, but to open channel for steamboats or lumber rafts, a line, running along tips of fins, was slackened and fins folded inward, leaving current to push boom towards shore. Line was managed by windlass seen at right.

Falls, again clearing the mill race of all obstacles, piers and booms and part of the sawmill. Millions of feet of sawed lumber piled in a nearby yard on the right bank were swept away like matchsticks. The mill had been sawing about 100,000 feet of lumber products a day, and grade one lumber was selling at $20 a thousand, the highest ever. The losses were staggering.

Repairs were made to the works, but production lagged.

On March 7, 1857, by an act of the legislature, Messrs. H.S. Allen, Jacob Willis, Elias E. Galloway, Eugene Shine, Otis Hoyt, their associates and successors, were "constituted and made a body politic by the name and style of Chippewa Falls Lumber Company with capital stock $500,000 divided into 10,000 shares of $50 each."

But the financial panic of 1857 was moving across the nation and its effects were soon felt in the logging industry and in the mills. Although Allen's new company failed, he refused to give up and clung to his machines and sawed lumber.

The big mill was handled by a succession of poor managers until 1862 when it was leased to Pound, Halbert & Co., who operated it successfully during 1863 and 1864. On April 26, 1865 at the close of the Civil War, the *Wisconsin Pinery* at Stevens Point published a report from a correspondent in Chippewa Falls which had this to say of the mill:

> Messrs. H.S. Allen & Co. have extensive saw mills on these falls, running five upright [saws] and two circular saws, besides a gang of fifteen more [saws]. Their daily product of lumber is 60,000 [board] feet.

Pound, Halbert & Co. bought the mill from Phelps & Corwin, trustees, on December 1, 1864. William Carson was associated with the company for a time in the late 1860s until 1869 when a new incorporation was organized under the firm name of Union Lumbering Company. Several new stockholders supplied additional capital which brought the total capital structure to $1,500,000. About $300,000 was used to improve the mill, and for new booms, piers and a dam. The capacity of the mill was boosted to 350,000 board feet in twenty-four hours and this made it one of the largest in Wisconsin.

Water inside this flume along west bank of Chippewa River carried logs from Dells dam in Eau Claire into tunnel which disembogued into Half Moon Lake. Opposite, upper and lower: different views of flume, the one above looks upstream, the one below, downstream. Catwalk on upper flume was used for men watching progress of logs.

Early view of Dells dam. At left, channel for logs entering flume to Half Moon Lake. Boom running at oblique angle was supported by jam piers to protect dam against loose logs and debris.

The Union Lumbering Company survived the financial panic of 1873, but two years later went into bankruptcy. The mill was released by the court to A.E. Pound and T.W. Halbert, who struggled with it for two years and also failed. The court-appointed receiver, L.C. Stanley, a local capitalist who also operated "Stanley House," kept the mill going until 1878, and, facing more debts than profits, he was forced to shut down. A year later William Wallace gained control of the mill and all the company timber for a sum of $150,000 and assumption of debts amounting to $300,000. Wallace leased the mill to a new firm, Beck & Barnard who, with other stockholders of the old Union Lumbering Company, organized Chippewa Lumber & Boom Company on December 19, 1879.

The new corporation operated the mill until the entire organization was sold on March 1, 1881, to the Weyerhaeuser Mississippi River Logging Company which turned it back to Chippewa Lumber & Boom. The price paid was $1,275,000 but this included the pine lands the company owned in northern Chippewa County and other counties.

In 1886, the big mill at Chippewa Falls was struck by lightning and burned with a loss of $75,000; it was rebuilt and the circular saws and bandsaws continued to whine day and night until the last pine log was sawed on August 2, 1911, and the mill was shut down forever and sold for scrap.

It is difficult at this point to understand why a mill, employing several hundred people, could not have made a profit for so many years. No doubt, the most important decision Weyerhaeuser made after he bought it was to appoint William Irvine as his production chief and general manager. Here was a man who had risen from the ranks in the lumber industry, even to studying the machinery in the sawmill. He visited the logging camps frequently and took a personal interst in the drives every spring. Although he mixed freely with the mill hands and men in the woods, he always maintained a rather stiff personal decorum, even to wearing the high starched collar and tie when visiting camp. His daughter, Mrs. Ruth Richter, recalls one time he was fishing for muskie, wearing a business suit, white collar and shirt and straw hat. He was standing in the stern of the boat casting, smoking a cigar when he got a big strike. The muskie unbalanced him and he fell overboard. The guide jumped in to help him back into the boat. He was still clinging to his pole, hat on, the muskie on the line, and cigar between his teeth!

Meanwhile, when H.S. Allen sold his mills on the Red Cedar, at least one, the Wilson Creek mill, was taken over by two young entrepreneurs from the east, John H. Knapp and Captain William Wilson. After operating the mill for several years, they took in a third partner, Henry Stout, who came with badly needed capital for expansion. Thus was born the triumvirate of one of the largest lumber companies in Wisconsin; by 1899, more pine was being sawed in Knapp, Stout mills than in any other in the state. The company was in business from 1856 to 1901. The water mill of Knapp, Stout in Menomonie sawed its last log on July 31, 1901, and the shingle mill closed August 8, 1901 and the last cribs of lumber went down the chute of the rafting shed on August 12, 1901.

At one of the last board of directors meetings the following partners and stockholders were present: W.W. Cassidy, J.E. Stout, J.H. Douglas, H.E. Knapp, L.S. Tainter, F.D. Stout, H.I. Stout, F.E. Wilson, J.H. Knapp, and T.B. Wilson.

While Jean Brunet was struggling with his mill in Chippewa Falls, there were other investors looking

Below: a later view of dam showing lock for lumber rafts and steamboats. Camera has caught a lumber raft (lower) emerging from lock.

Right: wooden draw bridge, completed in 1867 at Eau Claire, connected Kelsey Street (East Grand) and Bridge Street (West Grand). Drawbridge was for steamboats. View looks east.

for a good mill site, and one of these was Captain George Wales, an ex-army officer who had worked for Lockwood and Rolette on the Red Cedar. Wales found a site on the upper Eau Galle (pronounced Eau Galley) River, a feeder stream which runs into the Chippewa above Durand. Wales began sawing in his new mill some time in the winter of 1838-39, and not long after he took in two partners, a Captain Dix and Thomas Savage to form the Eau Galle Lumber Company.

Competition came when William Carson and Henry Eaton built another mill on the lower Eau Galle. This mill was built primarily to saw bolts for shingles, and to saw quality square timbers. Carson and Eaton, having located on the lower end of the Eau Galle, were in a position to obstruct or delay lumber cribs which were rafted past their mill from the Wales mill upstream. These obstructionist tactics ceased when the Eau Galle Company sold an interest in their mill to Carson and Eaton. Savage and Dix left the company, and finally Wales sold out, leaving the Eau Galle entirely in the control of Carson and Eaton.

In 1848, a second sawmill was built in Eau Claire, although the names of the builders seem to be lost. Nelson C. Chapman and Joseph C. Thorp took it over and this marked the founding of one of the giants in the lumber industry—the Eau Claire Lumber Company. In the next fifty years it sawed more lumber than any other on the Chippewa River although it was actually located on the north bank of the Eau Claire a short distance from the Chippewa.

By 1860, there were several mills in Eau Claire in addition to shingle mills, lath mills, planing mills. As these developments continued to expand, the owners of the big sawmill at Chippewa Falls realized they were being surpassed by the productive capacity of the mills in Eau Claire, and a fierce competition arose between the two cities, much of it counterproductive.

The Eau Claire mills stored much of their log supply in the Paint Creek dam and Eagle Rapids dam located above Chippewa Falls and there were times when mill owners from Eau Claire were having their logs delayed for one reason or another by the Chippewa Falls mill owners. The Eau Claire mill men wanted their own holding pond and they had a perfect place for it at the Dells above the city. In the winter of 1860 the mill owners asked the legislature to charter a company for the purpose of building a logging dam at Eau Claire. This was the legal route to follow, since no dam could be built on any stream in Wisconsin without a franchise from the state legislature, no matter how small or insignificant. This was one of the provisions implied in the Northwest Ordinance of 1887.

The state legislature approved the franchise but Governor Alexander W. Randall vetoed it on the grounds that it did not provide sufficient guarantees for a lock to aid steamboats and lumber rafts.

The Civil War delayed local construction projects nationwide and it was not until 1868 before Assemblyman Charles R. Gleason, representing Eau Claire and Pepin counties, pushed a bill through the legislature to charter a dam only to have an unfriendly presiding officer delay signing it long enough to kill it.

Efforts to get a dam were renewed in 1872 by Senator Joseph G. Thorp when another bill passed both houses only to have

Governor Lucius Fairchild veto it. In 1872, the state constitution was amended and the legislature was prohibited from granting corporate powers and privileges except to cities. This was Eau Claire's opportunity, but before any action could be taken, it would be necessary to bring the three villages of Eau Claire, East, West and South, under a city charter. This took time and it was not until 1875 that new legislation was introduced.

By this time, the mill owners in Chippewa Falls were visibly disturbed. Pound, Halbert & Company, which was operating the big mill under the firm name of Union Lumbering Company, published a descriptive map of the Chippewa River accompanied by some highly charged "remarks" directed at the mill owners in Eau Claire. The map of the river between Chippewa Falls and Jim Falls was drawn, ostensibly, to show how the company was going to improve log driving facilities on the river if they were only given a chance to do so. But the hostility reflected in the remarks made by the Union Lumbering Company publication suggests how deep the suspicion was between the two cities, each fearing sabotage from the other. The published remarks of the Union Lumbering Company follow, in part:

Here is presented the solution of the problem—How can *logging, manufacturing, and pine land* interests of the Chippewa Valley be secured and protected? For sixteen years the all important question with all parties identified with logging on the Chippewa River has been, what proper improvements in the river will best insure safety of logs, facilitate their delivery to the *mills*, and markets, and at the same time leave *unimpaired* and *unobstructed* that portion of the river which is practically navigable for rafts and steamboats!

Taking advantage of this great *public* necessity, a number of speculative schemers have at four annual sessions besought the Legislature of Wisconsin to authorize a corporation to build a dam sixteen feet high across the Chippewa River, change the channel, close the old one by piers, booms, etc. etc. at a point about 16 miles (by river) below the head of navigation [Chippewa Falls] below the great natural powers and below the most important established sawmills. This scheme was mis-named the 'Dells Improvement' and was projected not more to provide storage for certain local mills than to promote corner-lot speculations and other private and local interests by crippling if nor absolutely destroying all interests above that point—Eau Claire. So manifestly wrong, unjust and unconstitutional was this scheme, that it has required on the part of its devotees the most stupendous combinations of lobbying, accompanied by diabolical misrepresentation, malicious slander and open and defiant corruption to secure for it a respectable show of success. So persistently has this *mad folly* been pressed that its consideration has already cost the state many thousand dollars, bitterness and personal hate have been engendered throughout the entire state, until it has become the dread of every newly elected legislator.

In the second paragraph of these remarks, the Union Lumbering Company warns if the bill is reintroduced in the 1872 legislature that "the widows and orphans of Eau Claire will be taxed to raise a lobby and corruption fund."

The Chippewa Falls mill owners kept the pressure on for three more years and then went down to defeat. Meanwhile, on September 30, 1872, the Eagle Rapids, Flood, Dam & Boom Company was organized with stockholders represented from both Chippewa Falls and Eau Claire. The Eau Claire mill owners had hoped to improve the river in the vicinity of the Dells but temporarily gave in to the Eagle Rapids project. Mississippi River Logging Company was against any dam above Eau Claire or Chippewa Falls as possible traps to delay the company log drives.

But work went ahead on the Eagle Rapids dam, which was completed in 1873, only to be washed out the same year. By 1875, the new dam on the same site was nearing completion, and the boom company offered mill owners on the Mississippi to sort ten million feet of logs daily in flood stage for eight straight days and pass Beef Slough logs for 12-1/2 cents a thousand, instead of 35 cents which it was charging the mill in Chippewa Falls. Weyerhaeuser flatly refused, hinting he was not interested in paying for something he had been getting for nothing, all of which added further to the feud which had now grown up between the Chippewa River and Mississippi River mill owners.

The normal flow of the river, when it rose in the spring of 1873, aided by a log jam above the dam, smashed part of this new dam with the result that thousands of logs were lost. It was repaired and raised four feet, and this new construction held the force of the river until the spring of 1875, when a flood broke through a section of the dam a second time, leaving a big gap. Again thousands of logs were lost, and $300,000 worth of logs were stranded in sloughs. After that, the dam was abandoned.

Meanwhile, armed with new authority from the newly chartered city of Eau Claire, a bill was re-introduced in the legislature to build a dam at the Dells, but not necessarily for log driving purposes. The Eau Claire *Free Press* reported that 1875 was an exceptionally dry year and as a result, the citizens of Eau Claire had suffered from lack of fresh water, so much so that it was decided that if the people of the city couldn't have a log driving dam, at least they should have one to provide for a water works. Thus, with tongue in cheek, the Eau Claire mill men, in 1876, pushed an amendment to another bill, totally unrelated to logging, authorizing the city of Eau Claire, not the mill owners, to construct and maintain a dam at the Dells for the purpose of supplying the city with fresh water. After another hot fight in the legislature, the bill passed and the governor did not veto it.

In December of 1875 the Dells Improvement Company was organized with a stock capital of $100,000. Elected President was J.G. Thorp, Vice-President, Orrin Ingram, Secretary, C.A. Buffington, and Treasurer, D.C. Clark. At this point in time it seems rather remarkable that the Eau Claire mill owners had to resort to subterfuge to get their dam, and yet, they must have had just as much right to one as the mill owners in Chippewa Falls.

The Dells Improvement Company made a contract with the city to build the necessary waterworks, but made a few "improvements" including a sluice gate, a lock for steamers and rafts to pass through, booms and piers above the dam to control and sort the logs, and a canal that was to connect the river with Half Moon Lake. The terminal date for construction was set for April 15, 1878; the canal could wait until later.

The Dells dam was an impressive structure, 428 feet long, 108 feet wide at the base and capable of raising a head of water sixteen feet high. The last gate was completed and closed on March 16, 1878, and bets were taken around the city on how long it would take to fill the pond. Actually the time astonished everyone, including the engineers. With the great flow of spring freshets coming down, the pond was filled in eighteen hours, and backed up the water more than eleven kilometers. This eliminated any threat from Ash and Pine rapids which were now submerged. The pond had a storage capacity of more than 125 million feet of logs, the largest of any on the river or even Beef Slough. The dam also delivered energy to operate a waterworks for the city.

The lock on the dam was a boon to upriver lumbermen. Although it took longer for a raft to go through the lock than over the dam, there was no danger of breakage. In 1882, the big mill at Chippewa Falls ran ninety-two rafts to Reads Landing with a loss of only 135 bundles of lath and four bundles of shingles which were carried as "top load."

The race or canal from Dells pond to Half Moon Lake was built to run logs into Half Moon Lake for storage, and also to supply logs for mills located on the lake. The race, which was a mile and two-tenths long, left Dells dam at its west end and ran parallel to the river for most of its length, and then turned west between Randall and Beech streets into a tunnel 1200 feet long to reach Half Moon Lake. The race was built of planks, and sluiced four to five feet of water. (See accompanying photographs.) First used in 1880, it was an instant success.

The Dells Improvement Company charged a toll on all logs, lumber rafts and steamboats passing through or over the dam which brought in about $50,000 annually for upkeep. In a normal season in the 1880s and early 1890s the company handled ten million board feet of logs a day, and 100 million feet passed through the canal to Half Moon Lake each year.

The new flume and tunnel which opened in 1880 to traffic into Half Moon was not the first to connect the Chippewa River and the lake. In 1857 a cut-off was dug between the river and the lake directly west of 10th Avenue on the West Side. A fin sheer boom, after it was invented in 1861, was hinged to a point about opposite 9th Avenue on the east bank of the river which bounced or shunted the logs into the channel.

Efforts to create an artificial stage of water for the cut-off were never entirely successful and explains, in part, why the lumbermen were anxious to build a flume and tunnel above the new Dells dam.

Meanwhile, after the dam was completed, one of the first projects to be undertaken was still another cut-off across a spit of land where the river arched its back and almost touched shoulders, north of Dells Pond. Logs coming down the Chippewa for Eau Claire mills went into the big storage pond, but logs destined for downriver mills were routed across the cut-off to avoid the holding pond. (See accompanying map.)

An important collateral of sawing lumber in the mills of Eau Claire, Menomonie and Chippewa Falls, was the manufacture of shingles. In the 1880s the shingle mill of Knapp, Stout & Company was probably the largest in northern Wisconsin.

John Gilmore (above), came to Chippewa Valley accompanied by son Albert, probably in 1869. He worked in woods in 1870s, skidding with oxen, and in 1880s and '90s, he ran own camps as logging contractor. Gillmore Creek (name is variously spelled as Gilmore or as Gillmore) in Ashland County is named after one of his camps. Photo above was taken in studio of W. L. Bachelders of Durand.

William Carson (above), pioneer to Eau Claire. Carson, born in Inverness, Canada, came to Chippewa Valley in 1838 and worked in Eau Galle. In 1874 he moved to Eau Claire and later became president of Empire Lumber Company at 1106 Menomonie Street. He donated land to city for park named after him. His wife Mary E. Smith (above) was born in Rutland, Vermont.

Fairly typical of homes Eau Claire lumbermen built is this one in 1500 block on State Street where B. A. Buffington and family lived. At right, carriage house and horse barn.

A man, his house and his mill. William Irvinc was born in Mt. Carroll, Illinois, and came to Chippewa Falls as a youth in 1865 to work at various jobs associated with lumbering. In 1885 he was made general manager of Chippewa Lumber & Boom Company, a Weyerhaeuser subsidiary, and continued to hold this post until last pine log of company was sawed in 1911 and big mill (below) shut down. In his will, Irvine donated land along Duncan Creek to city of Chippewa Falls for park named after him, and he also gave an endowment of $100,000 for upkeep. He married Adelaide M. Beardsley on October 8, 1873. He died December 26, 1927, seventy-six years old. At left is photo taken in 1981 of one time Irvine residence at 606 Superior Street.

CHIPPEWA
RIVER
BIRD'S EYE VIEW OF THE CITY OF
CHIPPEWA FALLS
CHIPPEWA COUNTY, WIS. 1874
1. Court House
2. Public School
3. Congregational Church
4. Methodist Episcopal Church
5. Roman Catholic Church
6. Scandinavian Lutheran Church
7. Episcopal Church
8. City Cemetery
9. Catholic Cemetery
10. Race Course
11. Engine House an
12. Tremont House
13. Central House
14. Wateman House
15. Scandinavian Ho
16. French Hotel
17. German Hotel
18. Wisconsin House
19. Union Lumberin

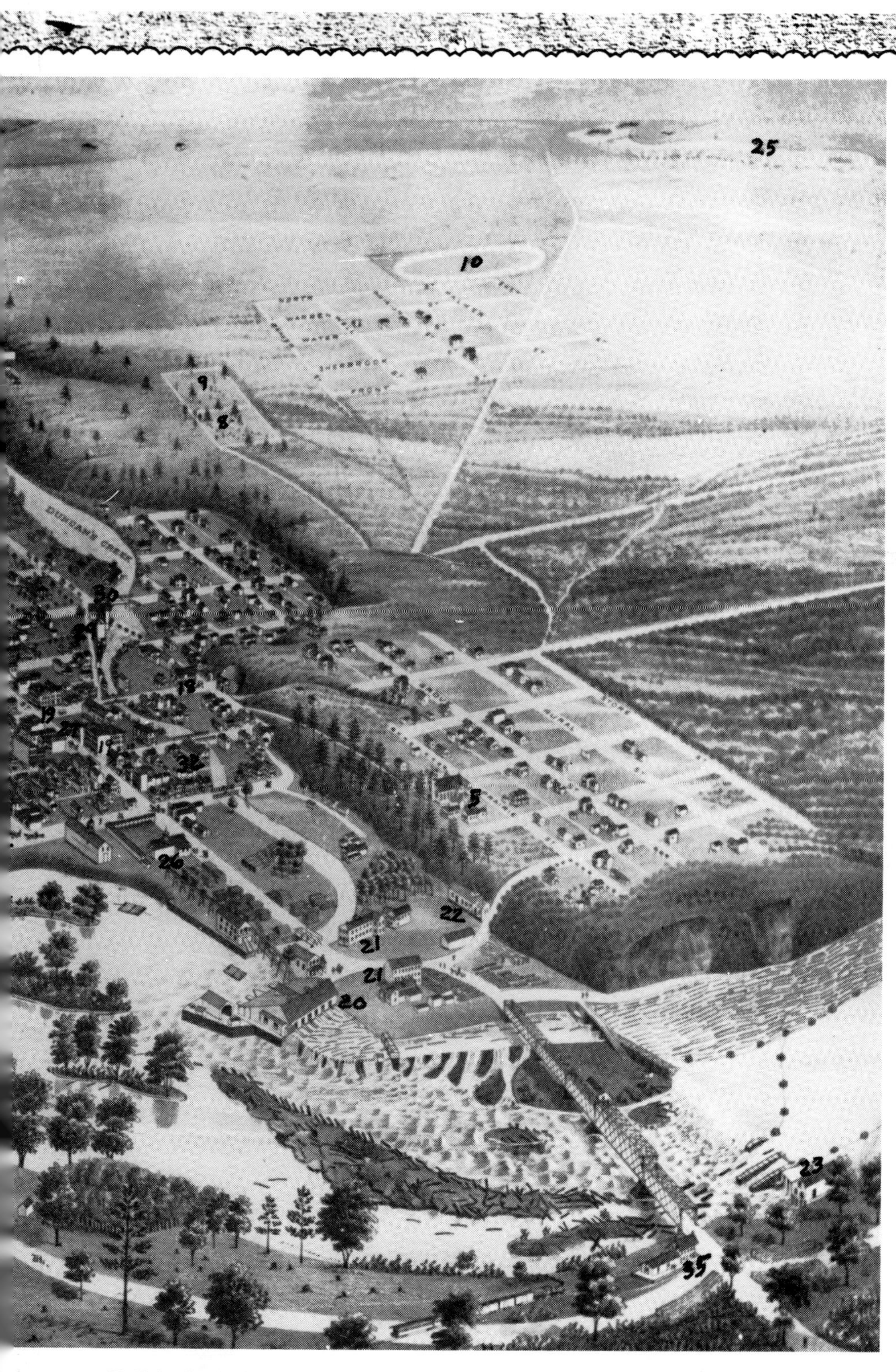

1)	20. Union Lumbering Company's Saw Mill	29. H. S. Allen's Flouring Mills
	21. Union Lumbering Company's Boarding Houses	30. H. S. Allen's Planing Mill & Sash, Door & Blind Factory
	22. Union Lumbering Company's Foundry	31. McRae and Witherow's Door, Sash & Blind Factory
	23. Union Lumbering Company's Shingle Mill	32. Styles & Co's Livery Stables
	24. Union Lumbering Company's Flouring Mill	33. Blair & Morse's Livery Stables
	25. Eagle Rapids Boom	34. French Town Steam Saw Mill
	26. Union Lumbering Company's Stables	35. Railroad Depot
	27. First National Bank	36. Leinenkugel & Miller's Brewery
s Office	28. D. E. Seymour's Bank	37. Schmidmayer's Brewery

View from hill looking southeast across building complex of Knapp, Stout & Company of Menomonie. At left is shingle mill. Pier runs from steam sawmill at right, and jam piers protect pier against loose logs.

In addition to big sawmill operated by water turbines, and second sawmill operated on steam, Knapp, Stout & Company of Menomonie operated what was probably largest shingle mill in northern Wisconsin. Pond monkeys with peavies in foreground are pushing logs under platform over to bull-slide where one log is already ascending to second deck of mill to be sawed. Mill was built directly over mouth of Wilson Creek. Shingles were shipped out by train but also as "top load" on lumber rafts en route to Mississippi River ports.

Another view from hill overlooking Wilson Creek on Red Cedar River (today Lake Menomin). In center is company office building. At extreme right, corner of shingle mill roof.

Close-up of dam which furnished power for "water mill" seen at left for Knapp, Stout. Gates on dam at right were equipped with "Tainter gates" invented in 1880 by Jeremiah Burnham Tainter, one of partners in Knapp, Stout & Company. Tainter gate was built like rocker and could be raised and lowered much easier than traditional rafter gate.

COUDERAY
Site of Upper Chippewa Valley's first sawmill; erected by Levi Felton and P. M. Parker in 1902.
In winter all supplies and equipment were hauled overland a distance of 40 miles from the nearest Soo Line.
In 1903 A.O. Nustad built the second sawmill at Eddy Creek one mile north on Reservation.
When government surveyors reached here in 1856, pine in this locality had been logged and removed by trespassers.
Eddy Creek Mill

Julliot Brothers sawmill at Seeley, Wisconsin. Standing on deck at left is William Julliot, engineer, and wife Mae. In center, hand on lever, is John, head sawyer, and leaning on canthook at right is Fred, carriage rider. Julliot Brothers had mill near Pepin in early 1900s doing custom sawing when they got contract with Pierce & Cable Lumber Company to move to Seeley with portable mill and saw lumber.

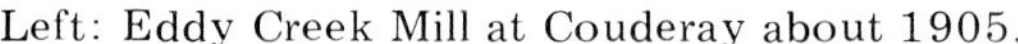

Left: Eddy Creek Mill at Couderay about 1905.

Although space was precious, one reason for high lumber piles was greater exposure to air and sunlight for drying. Pile at left appears to be two-inch planks. Horse is being used to jack planks to top. Source of picture uncertain.

Mill Sites Remembered

Without a careful study of abstracts and land titles, it would be difficult to pinpoint the exact site of all the sawmills on the Chippewa River, but from external evidence, news reports, engravings and plat maps, it is possible to locate most of them. In the accompanying sketch, courtesy of the Eau Claire *Leader-Telegram*, Memorial Edition, published in 1976, the mill sites are identified by numbers. Not shown on the map are four sawmill sites farther up the Chippewa River, one at the foot of Jim Falls built by Adin Randall in 1861. He later sold out to French, Leonard & Company, but the mill burned in 1874 and was not rebuilt.

New headstone which replaced earlier one dedicated to Jean Brunet in Forest Hills Cemetery, Chippewa Falls.

A century ago there was a community called "Chippewa City" located a short distance up the river from Chippewa Falls on the west bank where a mill was built about 1850 by Henry and Alexander O'Neill. The mill was actually on a creek soon to be called O'Neill, no doubt after the two brothers. The firm was sold to B.F. Branahan and A.C. Fair, and before it shut down in the early 1880s, it was owned by the Stanley brothers who ran a hotel in Chippewa Falls.

A third mill above Chippewa Falls was built by N.P. Bateman in the early 1850s near the mouth of Paint Creek, but it probably failed in the panic of 1857.

The fourth mill not shown on the map was located near the mouth of the Yellow River in the early 1850s, and later taken over by Gilbert Hedge Company of Burlington, Iowa. The fate of this mill has not been determined.

Point No. 5 on the map is the mill site of Joe Duncan on a creek, named after him, about three kilometers above the mouth on the Chippewa. The old dam site is still visible in Irvine Park. There were other owners of this mill, including George Warren and Daniel McHab, and finally Dr. W.T. Galloway of Eau Claire. The mill burned in 1871 and was rebuilt by S.M. Newton in 1874, and later A.E. Pound ran the mill until the company failed in 1877.

No. 6 on the map identifies the site of the mill built by Jean Brunet and, as already mentioned, by H.S. Allen and partners, and finally by Chippewa Lumber & Boom when it allegedly became the largest sawmill in the world under one roof.

"Frenchtown" was a community on the south bank of the Chippewa River opposite Chippewa Falls. Here a mill was operated in 1868 by James Mitchell and Edward Coleman. It changed hands several times before it was destroyed in the great flood of 1880.

No. 8 on the map identifies the so-called Gravel Island mill built in 1857 by Martin Daniels and Ephraim Shaw, both of Eau Claire. Problems with log storage and booms forced the mill to shut down.

No. 9 identifies the site of the first sawmill,

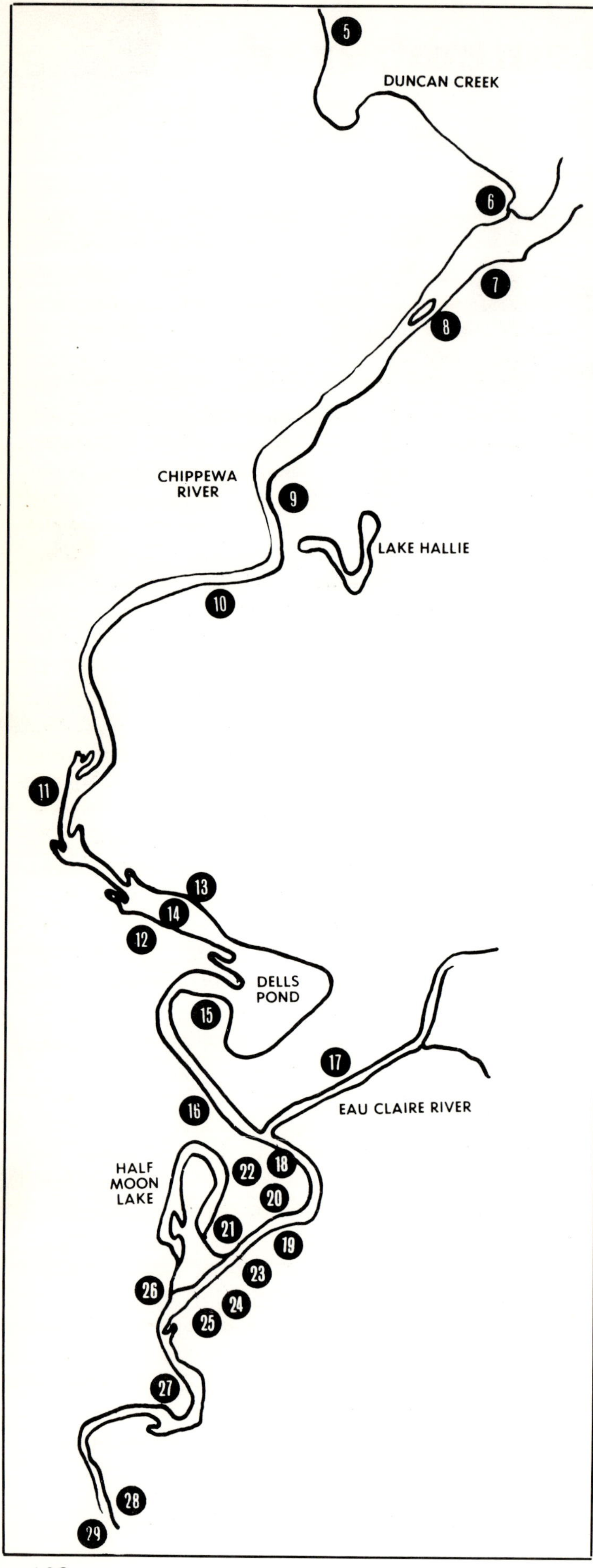

called the "Blue Mills," built in 1842 or 1843 north of the future city of Eau Claire by Stephen McCann and the Randall brothers. Later H.C. Williams and John Barron ran it under the name of "Badger Mills" and Williams apparently sold his interest in it to Barron who in turn disposed of his interest to Badger State Lumber Company. The 1888 plat of Chippewa County shows Badger State Lumber Company on the east bank of the river directly north of Lake Hallie, which corresponds with the accompanying sketch.

No. 10 identifies the "LaFayette" mill which stood about three kilometers below the upper Delles. It was built in 1863 by Charles Coleman and sold to H.F. Hodgins and John Robson in 1866. The mill had other owners, and finally ceased operations after damages sustained in the flood of 1880. The mill operated on steam in 1869, and a boiler blew up killing one of the engineers and injuring others.

No. 11 identifies the Wheaton mill built by Ira Mead in 1869. This, too, was run on steam power and cut five million feet of lumber a year, suggesting only one or two saws in operation. In 1878, the mill was taken over by the Northwestern Lumber Company and washed out in the flood of 1880 and was never rebuilt.

Leonard Farwell built a mill at No. 12 in the mid-1860s, but little is known about it.

The Barron brothers, John and Martin, built a mill at No. 13 across the Chippewa River from current Riverside Drive. The brothers took in a third partner, H. Clay Williams, and the mill later became the Nelson Hunter Company, but did not survive much beyond 1871.

Prescott, Burditt & Company built a mill at No. 14 about 1867, and enlarged it in 1873-74. The company was reorganized in 1879 under a new firm called Dells Lumber Company which operated the mill until 1898.

At the top of the loop in the Chippewa River, No. 15, north of the city, stood the "Big Eddy mill" built probably in 1863 by Arthur and John S. Sherman, both veterans of the Civil War. They sold the mill to the Ingram, Kennedy Company in 1866. In 1881, the latter company merged with Charles Horton Lumber Company, a big retail yard in Winona, Minnesota, and Dulany & McVeigh of Hannibal, Missouri, to form the Empire Lumber Company. The company also operated two mills in the Fourth Ward although neither is identified on the Index Map of Eau Claire published in 1888.

No. 16 on the map is attributed to the Pioneers Lumber Company. A birdseye engraving of Eau Claire, published in 1880, identifies the "Wilson Mill" in the lower part of the West Side on the west bank of the outlet from Half Moon Lake, with bull slide running into the Chippewa River. The Pioneers Lumber Company was organized by R.F. Wilson and his partners. It burned in 1883, was rebuilt with help from the Eau Claire Manufacturing Company, and was seriously damaged in the flood of 1884. It may be that the first mill built by Wilson was located at No. 16, but it was relocated to the site identified by the engraving after the fire of 1883.

The largest sawmill enterprise in Eau Claire was located on the north bank of the Eau Claire River near the Dewey Street bridge. No. 17 agrees with the engraving of 1880. There were actually two dams and two mills on the river, the one near the Dewey Street Bridge, and a second one a short distance upstream. The caption below the engraving of 1880 refers to the main mill as a "water power saw mill." This is an error. It began as a water power sawmill but had early converted to steam. The mill at the second dam was actually a water power mill and hence the confusion. Weyerhaeuser and associates purchased the entire plant of the Eau Claire Lumber Company and its new name became Mississippi River Logging Company. The Index Map of 1888 so identifies it.

No. 18 identifies the C.F. Mayhew Lumber Company located on the west bank of the Chippewa roughly opposite the mouth of the Eau Claire, but it is not identified on the engraving of 1880 because it was destroyed by fire in 1872.

In 1866, L.W. Pond and John C. Rose built a mill at No. 19 on the east bank of the Chippewa almost opposite the Mayhew mill, but, like the Mayhew mill, it was destroyed by fire, in 1870.

No. 20 on the West Side was the site of a mill which burned in 1875. Stephen Marston, a furniture store man in Eau Claire, purchased an interest in the mill from Adin Randall, and later sold out to Ingram, Kennedy Company.

Adin Randall built a mill on the site of No. 21 in 1856 and it was later taken over by Philander Ball and Frederick Smith. It probably stood in the southwest corner of Menominee and Ninth streets. Ball sold his interest to George Buffington, while William Carson acquired the Smith interest. The company became the Valley Lumber Company although the engraving of 1880 identifies this as the "Smith & Buffington" mill.

No. 22 refers to a mill site of the Sherman Lumber Company, or to the West Eau Claire Lumber Company, both on Half Moon Lake, the latter on the east bank and Sherman on the northeast bank of the lake. (See accompanying photograph.) The Shermans probably built their mill shortly after the canal from the river was opened in 1880, while the West Eau Claire Lumber Company was built in 1889. Both lost money.

No. 23 probably lay to the east of the Smith-Buffington mill on the Chippewa. It was built by the Chippewa Lumber Company organized in 1867 by S.W. Bangs, R.F. Wilson and Ira Mead.

No. 24 locates the big sawmill and planing mill of Ingram, Kennedy & Company, a stone's throw west of the Wilson mill between Menominee Avenue and the Chippewa River. The engraving of 1880 identifies it here, too. As already mentioned, this company was to merge with two others to form Empire Lumber Company.

No. 25 represented a small mill built in the early 1860s by John P. Pinkum. It probably sawed about 30,000 feet a day and was shut down in the panic of 1873.

At the elbow of the Chippewa River where it turns south below Carson Park stood the big mill of the Daniel Shaw Lumber Company identified as No. 26 on the map. The Shaw company owned a large tract of land for yards which extended east across the river. The Index Map of 1888 places the planing mill at the foot of the south entrance to Carson Park, but the location of the sawmill is not clear from the map. C.A. Bullen was an early partner of Shaw and when the latter died in 1881, he succeeded him as president. The company, one of the most successful in Eau Claire, survived into the

20th Century and sawed its last log on October 25, 1912. A year later, the mill and machinery were taken over by the firm of John H. Kaiser Lumber Company. Kaiser was cutting mostly hardwood, especially in eastern Sawyer County along the Brunet and Thornapple rivers. The mill ceased operations in 1924.

South of the Shaw Lumber Company premises, but on the Chippewa River, is a mill identified as No. 17 which appears on the Index Map of 1888 as "Westville." The mill stood directly on the bank of the river and at this point faced east, but no photograph of it appears available. W.B. Eastabrook probably built the first mill on this site in 1866. It later came to be called the Westville Lumber Company organized in 1882 by Aloney, John and Ralph Rust, and their brother-in-law, Frank W. Gilchrist, and three local men, the Rusts' half brother W. A. Rust, their brother-in-law, John S. Owen, and John Riggs.

No. 28 was the site of Porter's Mills, or Porterville, a community on the Chippewa south of Eau Claire in the town of New Brunswick where Gilbert Porter, in 1866, built the first sawmill on this site. Two years later he was joined by Delos Moon and S.T. McKnight, owner of a retail yard in Hannibal, Missouri. In its heyday, the company operated a shingle mill, planing mill and sawmill. Operations ceased in 1902 and the machinery was moved to Stanley.

Finally, No. 29 was a site opposite Portersville mills on the Chippewa where a mill was built in 1866 by John Stillman and the brothers A.W. and L.F. Gorton. At the end of four years there were no profits and no buyers and the mill was shut down.

There were other mills farther downstream as well as upstream from Eau Claire which started up, sawed and died, either for lack of funds or good management, or they burned. In the following account of sawmills on the Chippewa, William Irvine, superintendent for Chippewa Lumber & Boom for many years, lists the mills in the opposite direction from the above account, as he remembered them:

> Coming up the Chippewa in my early days we found the first mill at Meridean, some 15 miles below Eau Claire; then the mill of Porter & Moon, about 5 miles below Eau Claire; next the Daniel Shaw Lumber Company at Shawtown. At West Eau Claire the first mill was that of Ingram & Kennedy, (later Empire Lumber Company), then Smith & Buffington (later Valley Lumber Company), and next the R.F. Wilson mill.
>
> Just up the Eau Claire River from its mouth was the mill of Chapman & Thorp (later Eau Claire Lumber Company). Above Eau Claire, Ingram & Kennedy had a mill at the 'Big Eddy' at the head of 'Dells' Rapids. A mile above was the mill of Nelson & Hunter. Just above the lower Dalles, the mill of Prescott & Burdick (later Dells Lumber Company), then the Wheaton mill of Saul & Lally Wheaton, then the Lafayette mill of Hodgins & Robson (later J. & G. Robson and Lafayette Lumber Company); then the Blue Mills of John Barron (later Badger State Lumber Company); at Gravel Island, a few miles below Chippewa Falls, Bussey & Taylor (later French Lumbering Company and State Lumber Company); at Frenchtown, opposite Chippewa Falls, James A. Mitchell; then the big mill at Chippewa Falls. Above Chippewa Falls, the H.T. Rumsey mill at Eagle Rapids; Lockhart & Manahan at Chippewa City; the Gilbert mill on Yellow River, which was built in the 1850s and operated until some time in the 1870s when it, together with all the company timber, was sold to the Union Lumbering Company of Chippewa Falls; French, Leonard & Company had a mill at Jim's Falls.

Irvine adds that at the height of saw milling on the Chippewa River the production of lumber was about 350 million feet per year, which did not include Knapp, Stout's operations on the Red Cedar.

The New Dells lumber company was the last to close, probably July 13, 1929.

Above: Daniel Shaw Lumber Company's shingle mill in Eau Claire, about 1890. Mill stood on south side of Menomonie street near Michigan but Chippewa River and bull-slide do not show in picture. Mill manufactured some lumber but mainly shingles for roofing and laths for plaster board. Dump carts at left and right carried sawdust away, or kindling wood to customers around city. In center, a lumber wagon. In distance at right looms sawmill with silo-like slab burner, often called the "hog."

Below, another view of John H. Kaiser Lumber Company mill in Eau Claire, this one taken about 1907. Menomonie street runs back of buildings beyond mill at left. Logs shown here appear to be mixed hemlock and hardwood. An inlet from river formed small pond for logs to soak in below bull-slide running from second deck of mill down to pond. Through center of picture, left to right, runs road or railway spur on which logs are being brought into yard. Slab conveyor runs up side of mill and dump truck catches slabs.

Above: engraving of Chippewa Falls from *The History of Northern Wisconsin* (1881) looking west up River Street. Sawmill of Chippewa Lumber & Boom, and yards at left.

Stanley Hotel in Chippewa Falls, northwest corner of Bridge and Mill streets (later Grand), was home to lumbermen and wives, parties and social events. Hotel was remodeled several times and this picture taken at zenith of its popularity, about 1900. Building later burned. Dog looks askance at strange device at left, no doubt one of first auto-mobiles in Chippewa Falls.

The Sawdust War

The eleven hour day and even the twelve hour day was long an accepted work pattern in American industry and it continued almost into the 20th Century. In the logging woods of the 19th as well as the 20th Century the work day was often twelve hours, depending on the time of year. The men worked from dawn to dusk and most of them were out of their bunks an hour before daylight lest a precious moment be lost while they were walking to work.

The sawmills operated mostly on an eleven hour day, and in an effort to break this pattern, workmen in the Eau Claire Lumber Company mills on the Eau Claire River, walked off their jobs early Monday morning, July 11, 1881, demanding a ten hour day. Mill owners were thunderstruck. There was no warning, no apparent unrest, yet the men had taken the unheard of step of walking off their jobs under a slogan of "ten hours or no sawdust." Newspaper accounts fail to mention that the workers might have been members of a union called the "Chippewa Valley Workingman's Association" which was organized July 9, 1879.

The men from the Eau Claire Lumber Company left their mills and marched to the West Side and by nightfall all but one of the mills were on strike, the one exception being the Sherman mill on Half Moon Lake. When the strikers approached this mill, John Sherman is alleged to have told them he was already on a ten hour day, presumably from the moment he heard the strikers approaching, and ended up by inviting them all to a free beer at a nearby saloon.

The Eau Claire newspapers were horrified, and when three of the strike leaders were arrested for breaking the peace—which they probably did not—the editors were jubilant but not assured. They wanted the governor of Wisconsin, William E. Smith, to come to Eau Claire and see how low "Sawdust City" had fallen. He arrived two or three days after the walkout and after listening to the mill owners, and the strikers, he decided to call up several companies of the National Guard, all from southern Wisconsin. They arrived by train on July 23rd, and bivouacked on the grounds around the old Court House on the West Side. Officers established a command post inside the Court House, all reminiscent of Civil War days.

The strikers, many, no doubt, veterans themselves, met their opposites in the Janesville Guards and Watertown Rifles, and invited them to a picnic one day which the wives of the strikers had prepared in the park. A good time was had by all, but it was the end of the strike, too. Most of the mills had managed to keep their boilers warm and some of the machinery running with replacements, and the strikers realized unless they returned to work they were going to lose their jobs. The Valley Lumber Company, three months later, reduced its workday to ten hours, and most of the other mills followed suite in 1882.

During or after this struggle an anonymous poet in the Norwegian-American community composed a bit of doggerel under the title, *Krigen i Eau Claire* (The War In Eau Claire) which lampoons the mill owners, the editor of the Eau Claire *Free Press*, the city mayor, and Governor Smith. It was not picked up in translation by any English-language publication.

The late Trygve M. Ager, Eau Claire correspondent for the Minneapolis *Tribune*, found a faded copy of this long-forgotten poem and wrote a feature story on it for the *Tribune* on December 28, 1966.

The poem is divided into fourteen stanzas and is sung to the tune of an old Norwegian emigrant song. The first stanza sets forth the purpose of the strike: "We all up and declared, 'Ten hours is what we want; 12 hours a day is too long to stand in a sawmill . . . for such low pay.'"

Without violence, the poet points out, thirteen mills were shut down for "weeks nearly two," while "we were shouting with one voice, 'Ten hours is what we want.'"

The next stanza suggests that the mill owners were so angry about the strike that they were "ready to jump out of their skins" while the mouthpiece of the mill owners, the Eau Claire *Free Press*, shouted "murder" and "fire," the "rabble is taking over the city."

Each stanza closes with a triple "hurrah" for someone including the editor of the *Free Press.* Triple hurrahs were also reserved for the police who began to arrest people, and for the mayor who "left his common sense" and swore in a "lot of loafers" as deputies.

As to the National Guard, one stanza says the guardsmen arrived "ready for bloody war" and each day got their ration of crackers, water and bread, rye-coffee, beans, potatoes and bones without meat." A triple hurrah for the brave soldiers who came to Eau Claire expecting to find it half in ashes and instead found it "resting peacefully in sawdust."

In the final stanzas, it seems clear that some of the strikers, especially the leaders, had lost their jobs, and for this the poet calls for more hurrahs to honor the mill owners in the event they ever came to their senses.

While this anonymous Norwegian was lampooning the mill owners, the soldiers, on their return to their stations, also found time to joke about their participation in the "Sawdust War" and it was not long before someone, probably from the Lake City Guards of Madison, added his version of it under the title of "The Sawdust Campaign."

Oh, I'm a Lake City guard, sir,
I belong to Company C,
Of the gallant Fourth Battalion
Of the Wisconsin National G.

On glory I was bent, sir,
For fighting I did not pant;
So, to Eau Claire I was sent, sir,
The strikers to supplant.

My buttons they were bright, sir,
And loaded twenty rounds;
My belt was far from light, sir,
As I stept to martial sounds.

The strikers we did meet, sir,
On the banks of the Eau Claire;
We fought them long and well, sir
Despite our villainous fare.

The sawdust campaign is o'er, sir,
The rioters are foiled;
And we've been ordered home, sir,
With uniforms all soiled.

Oh, war is most cruel, sir,
Specially a sawmill scare;
But we are always ready, sir
To go again to Eau Claire.

Old photographs of sawmills seldom reveal the presence of a sawdust pile, although it was a common sight after the turn of the 20th Century. Before 1900 there was no law that said a mill owner could not dump his sawdust directly into the nearest river or swamp. The mill owners excused this pollution by insisting that the sawdust preserved the embankments of the river from erosion and filled the sloughs where logs often became stranded, all of which, of course, was nonsense.

In the 19th Century pickets for fencing around private homes were common. These were a byproduct of the slab that usually fell off the saw carriage in the mill when the log was being trimmed. After the turn of the century, there was little call for pickets and the slabs not used for fuel in the engine room were sent up an elevator chute and dumped into a steel tower resembling a silo. This was called a "slab burner." It had a heavy screen on top to keep the sparks from escaping.

A great deal of sawdust was burned in the slab burner, but companies engaged in the ice business also used some of it. Blocks of ice from the local mill pond or river were harvested in winter and moved up on a slide into an ice house where the ice blocks were piled, one tier above the other, and surrounded on all interior walls and on top by a heavy blanket of sawdust. In summer all that was necessary to get at the ice was to shovel the sawdust off the top layer and chisel out the required number of chunks and cover up again.

Some sawdust was hauled away from the mills by farmers for cattle bedding, but shavings from the planing mill were preferred.

But how can anyone over seventy from northern Wisconsin ever forget the pungent aroma of the sawdust pile where youngsters played and jumped?

Lumber pile for Fountain-Campbell Lumber Company at Donald, east of Holcombe on Omaha spur. Lumber piled on slant to provide drainage. Company later had mill in Ladysmith.

The "Tan Bark" Peelers

Socrates, the Greek philosopher, was forced to drink the juice of the hemlock tree and died, but in our time the juice of the hemlock has been used to cure hides for the tanning industry. Peeling the bark of the hemlock was a business that began in the New England states in colonial times, and in the 19th Century moved west with the flow of immigration. One of the most important tanning centers in Wisconsin was located at Rib Lake in the east of the Chippewa Valley where the soil favored hemlock trees equally as much as pine trees.

Learning of the great concentration of hemlock growth in Taylor County, Fayette Shaw, a "State O'Mainer," moved to Rib Lake township and in 1891 founded the Rib Lake Tannery. Tanneries were built by other companies at Medford, Spencer, Phillips and Rice Lake.

Hemlock bark, more commonly abbreviated to "tan bark" (tannery bark), was harvested both by the big lumber companies and by independent farmers. Peeling hemlock trees began in early spring and continued until about the first week in July at which time all the juice in the tree had oozed out and it was no longer feasible to debark the tree.

The bark on the tree was cut in four foot lengths, and the first cut was made at the trunk (see photo opposite). After this initial piece had been girdled, the tree was sawed down, the limbs trimmed off, and the rest of the bark removed by men wielding both axes and spuds, the latter a tool with a short handle and a shovel nose which was used to pry off the bark. After the bark was peeled, it was stood up against the fallen tree and allowed to dry, and later hauled to the tannery. In Rib Lake township there was so much tan bark harvested in the early 1900s that a steam hauler was used to bring in the bark from outlying lumber company camps. Farmers hauled in their own tan bark, usually after the first snow when wide-bunked sleighs could be used to haul loads twice as large as any wagon.

At the factory, the bark was ground into sawdust and thrown into vats of water and allowed to leach to make tanic acid, the main ingredient necessary in the tanning process.

Tan bark, like fire wood, was sold by the cord. According to Robert Rusch, editor of *A Pictorial History of the Rib Lake Area* (1981), a cord of tan bark delivered at the factory, circa 1900, brought six dollars which, by the economic barometer of the time, was good money and farmers could peel many cords in a season.

In the 1880s, the peeled hemlock trees around Rib Lake were left to rot, but in the 1890s, a demand arose for hemlock lumber, and one of the biggest buyers was the coffin industry which demanded a cheap but sturdy grade of lumber which could be covered with cloth material.

Load of "tan bark" arriving at tannery in Phillips.

Above, Ed Peterson shows how hemlock tree was girdled and bark stripped with ax and spud before tree was sawed down and remainder of bark removed. Below: cord of tan bark and crew. Left, Charles and Ed Peterson, August Enander, Victor Larson, and Arvid Enander. Photo taken in Rib Lake township, Taylor County, in camp of Ole Peterson, about 1907.

Rafting Lumber to Market

The rivers of Wisconsin were used not only to drive logs downstream to sawmills, but to move sawed lumber in "cribs" which, when combined with more cribs, became known as rafts. These were "rafted" downstream by men who steered them with long oars located, not on the sides of the raft, but on the front and back. No attempt was made to propell them by rowing. The river current moved the rafts, not the oarsmen.

On the Chippewa River steamboats were also used to tow, that is, to push rafts from Eau Claire or Menomonie to Read's Landing on the Mississippi River where several Chippewa rafts were brought together to form one giant Mississippi raft.

The movement of these combined rafts downstream on the Chippewa or Wisconsin rivers was called "rafting," not driving. However, the verb "to raft" also meant to build or assemble a crib of loose boards, twelve to twenty-five tiers high, and binding them with four foot stakes called "grubstakes" or "grub pins." This lumber was then rafted to markets on the Mississippi River by crews that attended them all the way and finally sold them to retail lumber yards in Dubuque and Clinton, Iowa, Rock Island, Illinois, and St. Louis, Missouri, where the largest retail yards in the Middle West were located.

The heyday of rafting lasted from the 1840s into the early 1870s, when rafting on the Wisconsin, for example, was largely replaced by railroad transportation. But on the Chippewa, lumber continued to be rafted downriver for many years after the railroad came to the valley mainly because of the ease with which a raft of lumber could be towed, i.e. pushed, down the Chippewa into the Mississippi from Dunville or Eau Claire as opposed to the long and hazardous trip required for a raft from Wausau or Mosinee to reach the Mississippi on the Wisconsin River.

Most lumber rafts in the Chippewa Valley were assembled at Eau Claire. There were rafts assembled at Chippewa Falls and even farther upstream, but until a lock was installed at the Dells dam in Eau Claire, it was always a hazard to run a lumber raft of any size through the rapids at Chippewa Falls or the lower Dells at Eau Claire.

At Eau Claire the lumber was assembled into cribs in "rafting sheds" where the cribs, once assembled, could be slid into the Chippewa River by tilting the platform on which they were made. In some sheds there were "ways" on which the rafts slid into the river much as a seagoing ship is launched after the customary baptism by champagne.

The making of a crib of lumber on the Chippewa River was similar to the cribs on the Wisconsin; the main difference lay in their size. The Wisconsin raft, composed of three "rapids pieces," was, compared to the Chippewa raft, small. There were too many falls and dams on the Wisconsin to accommodate large rafts, but on the Chippewa, at least below Eau Claire, there were no rapids or falls to bother the raftsmen. No pictures have been found to show a lumber crib in the process of construction at the mill sheds or on the platforms from which they were launched, but John M. Russell (Menomonie) has furnished the accompanying sketch to suggest how it was done. The cribs rode deeply in the water with only about three or four tiers of lumber above the water line. The average crib which went into a raft was less than a meter in depth. The water in the Chippewa River was not deep enough to support a heavier crib.

Seven cribs coupled one behind the other made a "string" on the Chippewa River, and four strings coupled together side by side formed the standard Chippewa raft. Larger rafts were only built when the level of the river was especially favorable.

No picture has been found at present to illustrate how lumber cribs were constructed on rivers of Wisconsin. Sketch by John M. Russell suggests how it was done on Chippewa River. Cribs were thirty-two feet long, and seven cribs coupled in tandem made a "string." Four strings coupled side by side made "Chippewa raft." There are three Chippewa rafts in picture below. Shingles, lath and pickets (for fencing) were carried on top of raft en route to Mississippi River markets. This picture probably taken at mouth of Red Cadar at Dunnville.

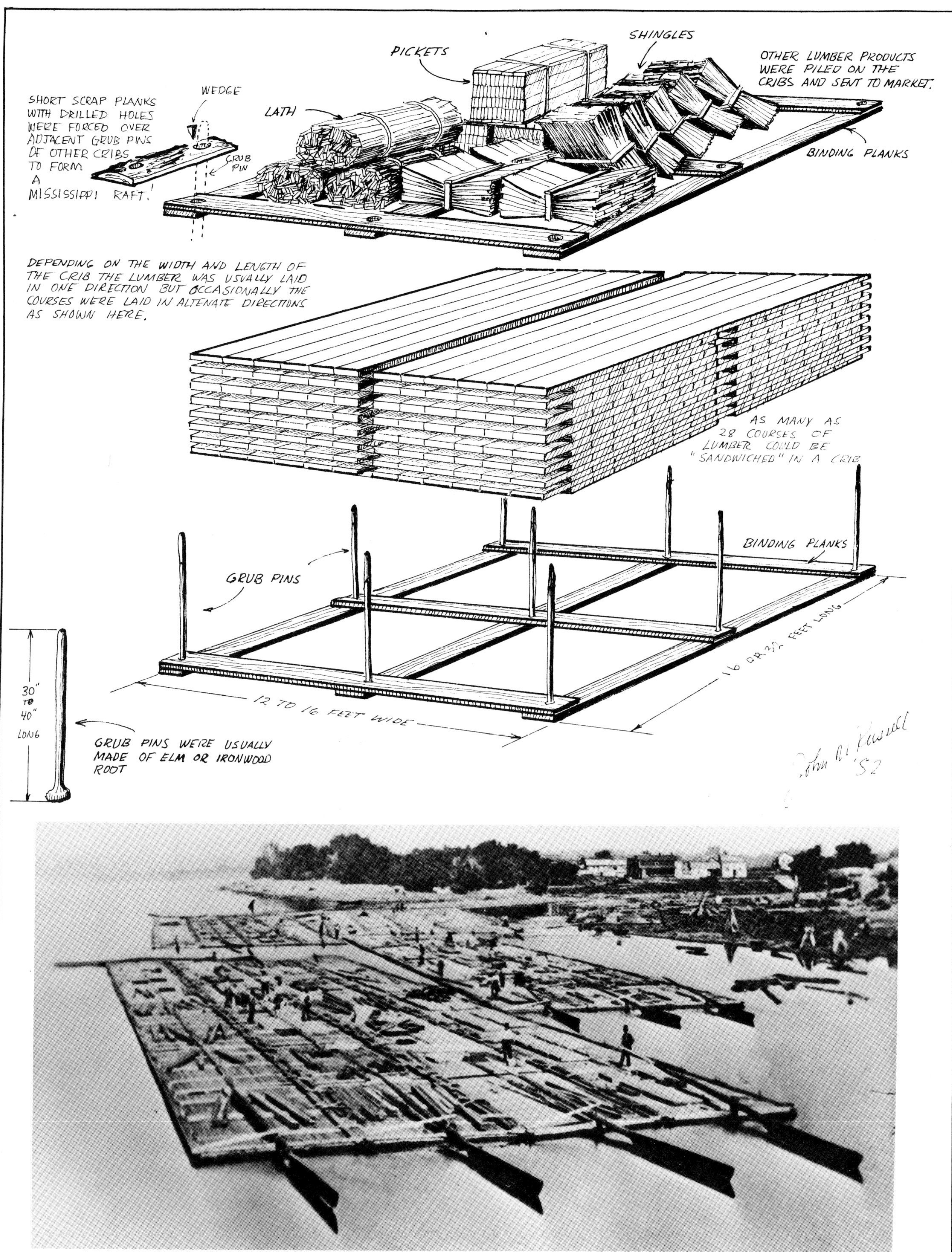
SHINGLES
PICKETS
OTHER LUMBER PRODUCTS WERE PILED ON THE CRIBS AND SENT TO MARKET.
WEDGE
SHORT SCRAP PLANKS WITH DRILLED HOLES WERE FORCED OVER ADJACENT GRUB PINS OF OTHER CRIBS TO FORM A MISSISSIPPI RAFT.
LATH
GRUB PIN
BINDING PLANKS
DEPENDING ON THE WIDTH AND LENGTH OF THE CRIB THE LUMBER WAS USUALLY LAID IN ONE DIRECTION BUT OCCASIONALLY THE COURSES WERE LAID IN ALTENATE DIRECTIONS AS SHOWN HERE.
AS MANY AS 28 COURSES OF LUMBER COULD BE "SANDWICHED" IN A CRIB
BINDING PLANKS
GRUB PINS
30" TO 40" LONG
12 TO 16 FEET WIDE
16 OR 32 FEET LONG
GRUB PINS WERE USUALLY MADE OF ELM OR IRONWOOD ROOT
John M Russell '82

For sales purposes, the solid contents of the standard raft of seven cribs long—a string—and four strings wide, were figured at 16 inches X 82 inches X the number of tiers of lumber, less ten percent.

A description of a crib of lumber being assembled into a raft is provided by Professor Charles E. Twining who writes as follows:

> The first step in constructing the crib was to lay the frame. Four 16 foot planks were placed lengthwise on the platform, two on each side, forming the bottom or underside of the crib. These planks now combined into two 32 foot sections, were called runners. Through the two-inch runners holes were drilled into which the grub pins were fitted. The iron wood or oak grub pins, shaped like plain ordinary pins, but four feet in length, fitted neatly into the prepared underside of the runners so that a smooth surface remained. The frame was completed

Scene of peace and beauty taken in dying sunlight at Dunville on Red Cedar River. Below mouth of river was works where lumber cribs for Knapp, Stout & Company of Menomonie were coupled into "Chippewa rafts." In foreground are makings of two "strings." Note grub pins, also called grub stakes, which bind tiers of lumber together. Crib of lumber rode deep in water, and long oars were used not for rowing but for steering. In late 1870s Knapp, Stout & Company began to use steamboats to push rafts down to Mississippi River and beyond. Note shingle bundles riding as "deck load" or "top load." Building at left is company store where there was dance hall on second floor. Building in center is boarding house for crew, and at right, probably horse barn. Henry Stout probably took picture with his new camera after discarding an early Kodak which shot picture with round frame. This picture is dated 1897.

with the placing of three additional 16 foot planks between the runners at either end and in the middle resulting in final dimensions for the crib of 16 by 32 feet. The lumber was then laid across this frame, perpendicular to the runners, forming the bottom of the crib. The next layer of lumber was placed lengthwise and the following layer crosswise, alternating in that manner until the crib was completed. Each layer or tier of lumber was called a course and the completed crib normally consisted of sixteen courses, although it often varied from twelve to twenty courses, depending upon the stage of water.

When the desired number of courses, or tiers of lumber had been reached, planks, complete with holes to fit over the grub pins, were placed down across the top layer of lumber. These planks, called "binders," formed the top frame in the same manner that the runners formed the bottom, and together the binders and runners connecting the grub pins, held the lumber together. The grub pins projected through the entire crib with a short piece at the top showing. To cap the operation, a long lever with a heavy iron clamp on one end, called a "witch," was placed over the end of each grub pin with a fulcrum bearing on the binder. By this means the grub pin was squeezed upward which caused the runner at the bottom and the binder on top to be drawn tightly together. Hardwood wedges were then driven into the grub pins from the top to keep them from slipping downward.

To steer the raft, a large oar lay fore and aft. This was made of pine about eight inches at the butt and thirty-four feet long. The blade of this oar was a plank sixteen feet long, twelve inches or so wide, and three inches thick where it was spliced to the stem, and one half inch thick at the extreme end. These long oars rested on headblocks at the stern and bow of the raft, and the oars were so evenly balanced, despite their length, that one man could handle an oar in normal water. But when passing over Paint Creek Rapids, two men were needed on each oar, and when passing the rapids at Chippewa Falls as many as five were required.

The standard Chippewa raft required a crew of ten or eleven men to guide it to Read's Landing. Eight were concerned with the oars, and the others served as pilot, cooks, and a man called a "snatcher," which will be explained below. Shipping the oars involved more pushing than pulling. The oarsman took the end of the oar stem in both hands and held it at arm's length above his head and walked or almost jumped with it across the deck of the raft with each stroke. This did not involve a steady push, but the rower seemed to surge forward only with every other step, and then,

Looking east across the Mississippi River from Minnesota side at Read's Landing to mouth of Chippewa River. In early period Chippewa rafts were assembled into Mississippi rafts at left, directly below Lake Pepin, but in 1870s and 1880s rafts were pushed across river to Read's Landing to be coupled into Mississippi rafts. Photo by author taken October 1981.

turning around quickly at the end of the sweep, he repeated the same steps back. Again, it must be emphasized, none of this motion involved propelling the raft; the oarsman could only steer it as best he could and leave it to the river to propell the raft forward.

When the long oars of the several men on board were not adequate to swing a raft around a bend in

time to prevent going aground, the "snatcher," carrying a "snatch pole" twelve to sixteen feet in length, went into action. The pole, tapered down to a steel point at the bottom, was circled three feet from the top by an iron band to which a strong ring was attached. Then a line, called the "snatch line," was secured to this ring and kept coiled at the stern of the raft. When approaching a sandbar or sharp curve, if the oarsmen could not maneuver fast enough, the snatcher grabbed his pole and jumped overboard while the pilot payed out the snatch line. The snatcher kept the sharp end of the pole plowing the bottom of the river while holding the other end at an angle of forty-five degrees, and the pilot snubbed the rope on one or more of the grub pins in the lumber cribs. This was done to slow down the raft in order to follow a course that would not have been possible under ordinary circumstances. After the raft slowed down sufficiently for the oarsmen to handle the situation, the snatcher pushed his pole into an upright position and removed it from the muddy bottom.

When the raftsmen reached Read's Landing, some of them turned around and went back to Eau Claire where more rafts were waiting. They followed a gig path on the left or east bank of the Chippewa about ten kilometers and then crossed the river, presumably in a ferry, to follow the right or west bank of the river to Durand. Many stopped over night at George Ecklor's hotel at Ella.

Although some of the crews returning to Menomonie or Eau Claire went by keel boat or by steamboat, others who walked had to cross the Mississippi from Read's Landing on the local ferry. This was merely a flatboat guided by a rope across the river which derived its power from lee boards which, when set at the right angle, propelled the boat across the river. The route followed by the raftsmen on foot as seen on the 1877 plat suggests that the ferry point would have to be located on the east side of the mouth of the Chippewa. In later years, a ferry ran between Wabasha and Nelson and there are many in the community who still remember it.

These raftsmen were all on salary and the lumber companies realized that the long walk back, at least two days, was costing them money. In 1871, Captain E.E. Heerman, skipper of the *Minnietta*, contracted with the lumber companies to bring their raftsmen back at $3.50 per head, although some wags said it cost him $2.95 just to feed a man. Signing this agreement with Heerman were the Eau Claire Lumber Company, Porter & Moon, Smith & Buffington, Daniel Shaw & Company, Ingram & Kennedy, Chippewa Lumber Company, John Barron, J.O. Nelson, Burdette & Company, Hodgins & Robson, J. Betch, Wheaton Lumber Company, E.W. Farwell, Wm. Estabrook, A. Boyd, John Rose, Frank Bradshaw, Union Lumbering Company, and Stanley Brothers of Chippewa Falls. Captain Heerman also agreed for this price to superintend the construction of brush-and-wing dams to aid in rafting lumber over shoals.

But a year later the newly built railroad to St. Paul on the Minnesota side of the Mississippi River made it possible for crews to return to Eau Claire via St. Paul on the West Wisconsin Railroad which came through from Eau Claire in 1870. For a time the lumbermen were torn between giving their business to Captain Heerman or the railway company. The captain also thought he had something coming for all the free time he had given pulling rafts off sandbars and bringing back raft kits—the tents, tools and kitchen ware each crew required. But Heerman saw the handwriting on the wall and found employment for his boats on the Mississippi that year.

The brush-and-wing dam mentioned above was a crude but effective way of raising the water level of the river at certain points. Brush was piled from the bank into the river and weighted down with stones, rocks, and deadheads. This created a barrier which diverted the river into a narrower channel and raised the level. A brush-and-wing dam extending out from both banks of the river could probably raise the river one meter. In China, the same principle is still used by sampans on the rivers in low water. Two heavy planks hinged to the mid-section of the sampan on either side, are extended outward, facing upstream, which raises the water under the boat and lifts it off the bar.

The practice of pushing a lumber raft with a steamboat on the Chippewa River apparently did not begin before the spring of 1879. The Pepin County *Courier* at Durand, in its March 23rd weekly edition, had this to report:

> Knapp, Stout & Co. have commenced a new style of running lumber. The Str. *Phil Scheckel* is now towing rafts from the mouth of the Menomonie [Red Cedar] River to Read's Landing. In this stage of water she handles a raft in fine shape and makes the trip in less than half the time it takes to float out the lumber. She has taken on three six-string and one seven-string rafts this week. The seven-string raft that went down this morning contained 645,120 feet of lumber with 190,750 shingles, 150,000 lath, and 25,350 pickets as top loading. This is believed the largest raft ever run out of the Chippewa.

Mention is made in the above to the "mouth of the Menomonie." Here a community developed called "Dunnville." Most of the men here were employed on the river coupling lumber cribs which were coming down the Red Cedar from Knapp, Stout rafting sheds in Menomonie.

The last lumber raft out of Eau Claire was run on June 29, 1901. This date was established in an unusual manner. Robert Donaldson, one of the raftsmen on this run, told Robert E. Boyd that he could not remember the precise date, but he did remember it was the night after "the elephant was killed by lightning." Searching the files of the Eau Claire *Leader*, Boyd found the story. "Stella," a trick elephant of the Wallace Circus, had been killed by lightning on the night of June 28, 1901!

William Irvine, superintendent for Chippewa Lumber & Boom, believed the last lumber was rafted out of Chippewa Falls in 1886.

At Read's Landing there was a works for assembling Chippewa lumber rafts into huge, almost unmanageable "Mississippi rafts" which were pushed downstream by stern wheel steamboats. There was no special technique required for assembling these big rafts. Short planks, called "coupling yokes," were lapped from the grub pins of one raft over to the grub pins of a second raft and these held the two rafts together which in turn were coupled to other rafts in the same manner.

Chancy Lamb, one of the Mississippi River mill owners associated with Weyerhaeuser, developed a double-spooled steam capstan which was built into most of the big steamboats after it was invented in 1874. With lines running from port and starboard of the steamboat to the stern corners of the raft,

A Mississippi raft passing down river en route to sawmills in Iowa, Illinois and Missouri. Stern wheel steamboat at left is pushing raft, and "bow boat" rides up front tied perpendicularly to raft to act as "point." By steaming ahead or astern, bowboat helped steer raft.

the steamboat capstan could reel in or pay out the lines to steer the raft after a fashion. The bow of the steamboat abutted on a block in the center of the raft astern, but from this position it could only push, not turn or twist the raft. Hence the advantage of the steam capstan, actually a winch.

In the bigger rafts, a second boat, called the "bowboat," was rigged across the bow and went ahead or astern according to signals from the steamboat astern. By the mid-1870s nearly all log brails were being pushed by steamboats, too. Piers and new bridges lower down the Mississippi were smashing the brails all too often and the invention of the double spooled steam winch was one tool in the lumbering industry which came in the nick of time unlike several other inventions, such as the jammer, which came into use almost at the end of the logging era.

Young people, no doubt children of Knapp-Stout-Wilson families of Menomonie, are on picnic cruise aboard lumber raft floating down Red Cedar River. Top-load in foreground mostly boxes and shingles. Although lumber cribs are yoked together with coupling yokes, lines at right have been tied to grub stakes (also called grub pins) as safety measure. In distance note erosion of bank on bend in river. Young man at left wears knickerbockers.

The Little Falls Dam

The community known in the 1880s and 1890s as "Little Falls" on the Chippewa River never had a post office until 1898 when the name "Martin" was given to it, with A. (for Amasa) J. Edminister serving as the first postmaster. Before that time, the mail was picked up at Estella.

The first plat of the village was laid out by Adolph Bernier, a local businessman, and recorded in 1902 as "the Plat of Little Falls." The Omaha railroad was coming up from the south that year and the company wanted to locate a depot nearby, but Bernier and the railroad authorities could not agree on terms. As a result, the railroad bought land to the south of the dam, platted its own village and called it "Holcombe" and the post office got its new name on December 6, 1902.

In the early years the community northeast of the dam was called "Barney town" after Bernier. After the big dam was abandoned most of the houses in Barney town were moved or demolished and Holcombe village developed to the south of the dam. Today, a tourist pausing on the shoulder of County Trunk M at the bridge may look north across Lake Holcombe, or, as some people still call it, the flowage. It is a beautiful body of water, spaced with islands, glinting and rippling on the quiet days, raging and foaming with white caps on the windy days. But all is safe and sound behind the big hydroelectric dam of the Northern States Power Company located a short distance below the bridge on M.

But without a map, there is no way of knowing how much more of the lake lies around the bend. Nor is there any way of knowing what an important piece of real estate this once was, and what it meant to people working in the sawmills of Chippewa Falls and Eau Claire, and to people in the Mississippi Valley who were begging for forest products to build their homes and barns.

Now, if the tourist could look at this through the dimension of history, this lake, which grew from the great river called the Chippewa, would leap into life, for this was once the beat of the river drivers, of floating logs, of booms and jam piers, all tied to one man-made phenomenon, namely, a wooden dam stretching from the east bank to the west bank of the river which was said to be the largest wooden dam in the world. It was so wide that a photographer who took a picture of it, some time before 1884, had to back off quite a distance to get both ends of the dam into his frame and even at that he cropped the west end. Altogether, it measured 625 feet, or more than the length of two football fields. Each of the thirty-two flood gates was seven feet wide and seventeen feet high. It could raise a "head" of water sixteen feet high. Most dams in Wisconsin, whether for driving logs or for driving the machinery in a sawmill, raised a head of eight to ten feet. But the Little Falls dam rose from foundations sixty-three feet wide, and abutments having a width of a hundred feet. In addition to the flood gates, there was a sluice gate for sluicing logs, and a smaller gate for running the wanigans through.

Below the dam a short distance the engineers also built a wing dam. This was probably a wall of timbers, buttressed by cribs filled with stone and gravel, which ran obliquely from shore downriver to force the river into a narrower, therefore faster channel, and to avoid an outcropping of rocks in the middle of the river. To date, no pictures have been found to show on which side of the river this wing dam stood but it was probably on the west bank. It either washed out in the flood of 1880 or some time before June 1882, since it is never mentioned in a daily journal kept at the dam beginning June 22, 1882.

Some years later a small wing dam was built from the west bank to the rock outcropping in the river, a distance of about sixty meters to channel the water away from the rock pile. James Jardine was foreman of the dam crew in the early 1900s and when operations ended in 1911, he was hired by the utility company, which had bought the property, to act as overseer and watchman. His son, Zac, remembers the second wing dam from his youth when he crossed over on it to fish. He said it was reinforced with cribs filled with rocks and buttressed by a phalanx of logs leaning against the deck at an angle of about forty-five degrees on the upstream side of the deck.

When Zac went off to war in November 1917, he remembers that the flowage above the Little Falls dam was still up. During World War I his

First dam built by Beef Slough Company at Little Falls (Holcombe) in 1878. Dam stretched 625 feet from shore to shore and was destroyed in flood of 1884. New dam was built on same site, but at different angle.

family moved to Madison and Zac lost track of events at the dam; oral tradition holds that a flood in the early 1920s washed out the last gates and also carried away the small wing dam.

About a thousand feet above the dam stood two rows of jam piers, alternating, one forward, one back, straight across the river. Early accounts also mention "ice breakers" but these were not evident at the turn of the century and no photograph has been found to suggest what they looked like. They were probably specially built jam piers which helped break up the ice flow in spring.

Directly in front, that is, on the upstream side of the dam, on either side of the river, were two long platform booms hinged to piling on shore and tapering downriver at an oblique angle toward the sluice gate. These booms acted as a funnel for incoming logs. The "pond monkeys," using their long pike poles, pushed the logs forward toward the sluice gate from the platform boom, and occasionally they jumped on a big log and rode it to a spot where the logs weren't moving fast.

When the gates of the dam were closed to build up a "head" or to "catch a head," as the expression was used, the dam created a reservoir that extended up river several miles almost to the mouth of the Flambeau River. The cost of the entire works, dam, booms, wing dam and jam piers was about $90,000 in 1878. It was built for this relatively small sum because wages were low, the planks and timbers were cut on land owned by the lumber company, and a local mill sawed most of it. There was also a stone quarry nearby, actually the rock outcropping in the middle of the river below the dam, a phenomenon quite visible on several old photographs.

The purpose of the dam was to control the water level on the Chippewa River all the way to its mouth on the Mississippi, or to Beef Slough. By controlling the flood gates and releasing the water when the logs in the pond were ready to be sluiced, the level of the river below could be raised to carry the logs across the sandbars, tow-heads and shallows all the way to Beef Slough.

In an effort to determine what the new dam was capable of, the gates were all shut down on November 8, 1878 at 9 a.m. which caused the river to fall at Eau Claire on the following day by one foot. Twenty gates were opened on the 13th, with a fifteen foot and ten-inch head of water, and allowed to run twenty-four hours, during which time the river level directly below the dam rose six feet, and at Eau Claire, four feet six inches, and at Durand three feet one inch.

On most rivers of Wisconsin there were usually two gates, one for sluicing logs, one as waste or flood gate. Cameron Dam on Thornapple River had three, and the Dells Dam on the Wolf had four.

Drawings by Randall E. Rohe shows how splash planks were inserted in front of dam at Little Falls whenever it was necessary to build up a "head" of water for sluicing. Lower drawings show how rope was inserted with eyebolt into one end of plank for removing it with aid of block on A-frame.

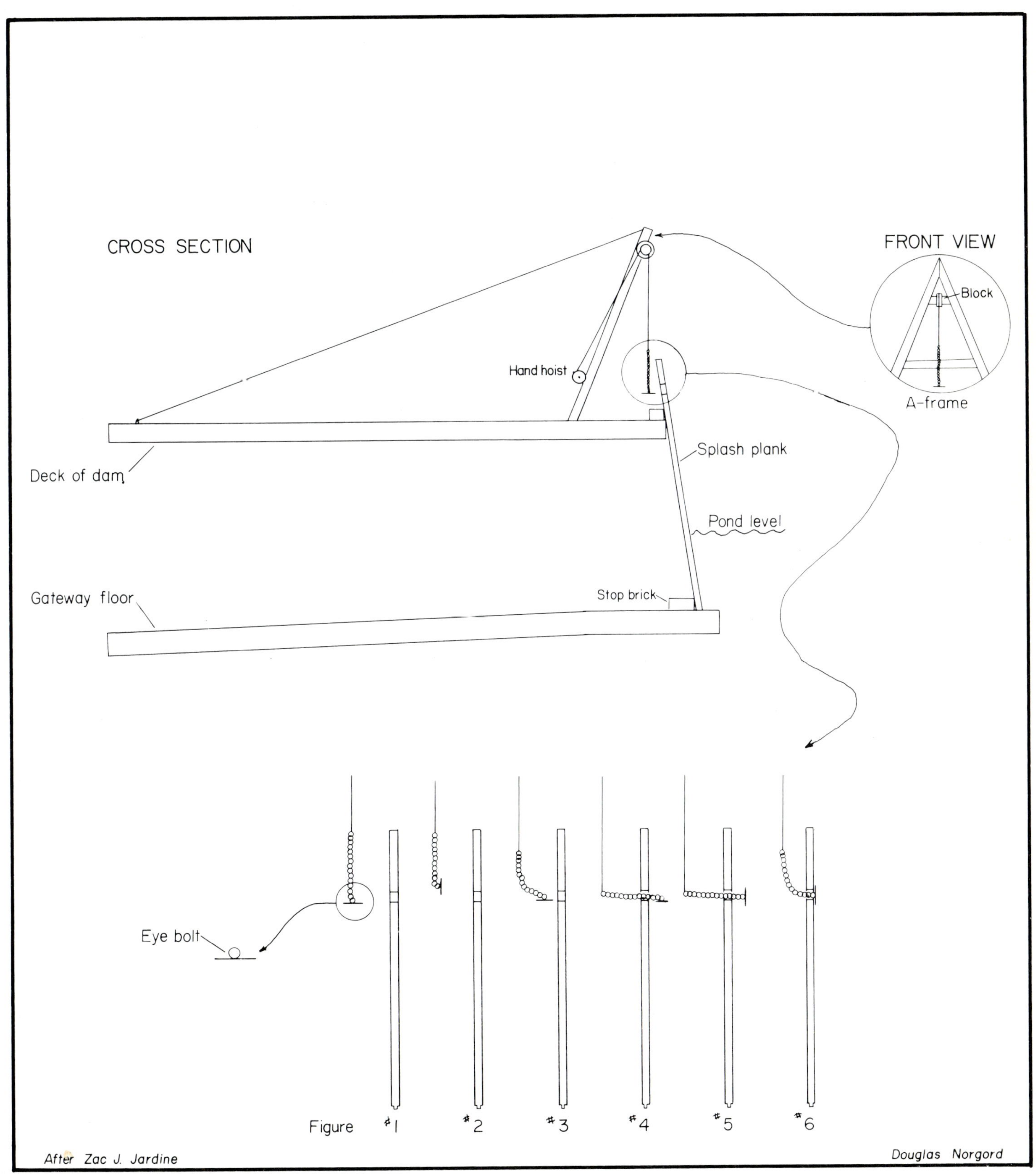

These gates were usually built with fender posts on two sides supporting a frame of several horizontal planks which could be raised by a windlass, all variations of the rafter dam.

But the dam engineers at Little Falls, Elijah Swift and Joseph Viles, turned things upside down, and instead of gates with horizontal cross pieces, they built gates with huge planks that stood on end, leaning against the dam at an angle on the upstream side. These were called "splash planks," in some texts referred to as "needles," and the entire dam as a "needle dam." The splash planks at Little Falls were about four inches thick, eight to fourteen inches in width, and in length not less than eighteen feet. The planks had to stand at least three feet above the deck of the dam they were leaning against. The splash planks, it must be repeated, stood on end, forming a phalanx of planks held in place by the pressure of the water.

Zac Jardine, who was born in the company house at Little Falls dam, remembers how the splash planks were handled. He said two men manipulated the planks with the aid of an A-frame which had a pulley at the top and a winch across the lower bars of the frame. The A-frame was supported by a guy rope secured to the back of the dam, resting against a wood railing running along the deck. A rope ran through the pulley down to the winch, and at the end of the rope was a piece of quarter inch chain, two or three feet long, which was fastened to an "I" bolt made of about one half inch rod and bent into an eye in the center. The chain was then secured to the eye, and the eyebolt on the chain was turned parallel to the chain and slipped through a hole augered through the upper end of the plank. The eyebolt was then turned at right angles to the chain to hold the plank. When the splash plank was ready to be lowered into the water, it was lined up next to the first plank and shoved into place. A third man assisted in this phase of the operation by guiding the plank into the water. The bottom of the planks rested on the bottom of the gate opening against a block of timber or offset running the width of the gate.

When the splash planks were pulled out of the water, they were laid alongside the plank road that extended across the length of the dam, a road wide enough for a team of horses to follow but not to meet another team. Until a steel bridge was built below the dam in 1906, the plank road was the only way a team could cross the Chippewa at this point. Part of the plank road had railings, and where there were no railings, the teamster usually got down from his wagon and led his horses across.

The first dam at Little Falls was severely tested in 1880, when a flood washed out the embankment on the east end. This released the water in the flowage and, together with the heavy rains, the river below the Falls became a torrent and logs poured through the breach, adding to the force of the flood. Thousands of logs were lost. In mid-June there was another flood, which raised the level of the water at the Dells dam in Eau Claire over the flood gates. This lifted the booms over the top of the piers and, with nothing to hold them, the booms sloshed over the dam followed by thousands of logs, sweeping everything before them.

The wagon bridge between east and west Eau Claire was carried away; Sherman's sawmill on the west side was carried off; the Schulenberg mill was badly damaged; the Daniel Shaw mill escaped undamaged, although the other mills on the lower West Side were undermined; rafting sheds, slides, booms and piers in front of the mills were swept downstream. Most of the logs passed through Durand and the fin sheer boom at the head of Beef Slough "worked like a charm," according to the Pepin County *Courier*, until the full force of the flood stage reached Beef Slough. At this time, the water rose over the piers, and the sheer boom floated downstream for some distance before it ran aground. Boom boss, Albert ("Ab") Gilmore, and several of his men, rode the boom and finally tied up on the "small tow-head below Deer Lake," said the *Courier*. Two river steamboats, the *Hartford* and *Artemus Lamb* towed the boom back to Round Hill where it was anchored once again. "The mill men who d--m the boom in low water," said the *Courier*, "will have to bless it in high water as without it they would have lost several millions of [board feet] logs which would have gone into the Mississippi and been entirely lost."

In the next several days mill owners in Eau

Claire sent crews out to retrieve the stranded logs which had drifted into the bottoms and woods, and company steamboats, including the *Artemus Lamb, the Iowa City, and Silas Wright,* were brought up the river to assist in picking up booms. *Artemus Lamb* was probably the largest steamboat ever to ascend the Chippewa River. On board were officials of Mississippi River Logging Company who came to view the devastation.

After the break was repaired on the east end of the Little Falls dam the entire dam was raised five feet to catch a head of twenty-one feet. This enlarged the flowage upstream but it also weakened the overall structure, and, when the next big flood struck in 1884, the added pressure on it made it more vulnerable than the original dam built in 1878.

The editor of the Pepin County *Courier* at Durand took a dim view of the Little Falls dam and all its works. In his weekly newspaper on Saturday, November 18, 1878, he wrote:

> No steamboat communication with the outside world this week until today when the *Monitor* came up [the Chippewa River] owing to the Little Falls dam being shut down. Is it not a little singular that the Government should lay out a large sum of money to improve the mouth of the Chippewa and then allow private corporations to put in dams and "improvements" that enable them to "dry up" the river at pleasure. It won't be but a year or two longer before the Chippewa will be reduced to a log driving stream and nothing more.

But the editor, W.H. Huntington, a former clerk on a river steamboat, had only that year taken over the paper and he probably did not realize what the Little Falls dam was going to mean to the economy of the entire Chippewa Valley. The steamboat traffic, which carried mostly freight such as farm implements and wheat out of Durand and Eau Claire, was small compared to the tens of thousands of pine logs which passed Durand every year en route to Beef Slough and the Mississippi.

The great dam at Little Falls served its builders well for the next four years from 1880 to 1884 and then disaster struck when another flood took out the center and most of the east end. The crises came on September 17, 1884, and this is how the keeper of the daily journal at the dam described that day:

> All gates up. Every piece of splash taken off and everything done to let water escape and still rising in pond. About 5 p.m. o'clock the dam went out, the far side bursting first about half an hour before this side. The logs piled up over the dam. Blocked up the gates and 'bust' the booms taking some of the piers with them.

By the far side, the journal means the east side.

The next week the crew was busy picking up the pieces. Roads were blocked with logs which floated ashore. Booms above the dam were smashed, part of the company warehouse was carried away, and the blacksmith shop and an old stable were gone.

Word was sent by telegraph to Frederick Weyerhaeuser and without any hesitation, it seems, he and his associates decided to rebuild the dam and word came back for the work to get under way. On September 24th this entry appears in the daily journal: "Men preparing to build dam," and by October 4th work had begun on the coffer dam. But the new dam was not going to be built as a copy of the old, that is, in a series of gates stretching straight across the river from bank to bank. Instead, it would be built at two angles, a zig and a zag, or, as Charles Henry, a boss on the drive, observed, "crooked like a dog's hind leg" to give more space for spillways and reinforcement.

William ("Billy-the-Beaver") England, one of the engineers of the new dam, had a reputation of "never having lost a dam." And he didn't lose this one either. Except for a gate or two that washed out in later years, it stood firm as a rock and served the lumbermen until the final drive in the summer of 1911.

The flood of September, 1884, destroyed not only a dam, but created havoc along the entire length of the Chippewa River. It even attracted national attention. *Harper's Weekly* ran a brief story and an engraving in its September 27th issue, and, as so often happened, confused board feet of logs with logs. The *Harper's* story reported that the flood:

Construction work in mid-winter on new dam after flood of 1884. Snowbank appears under first arch which will be a "bear trap" sluice gate. Iron track carried carts with construction materials, rock and timbers for piers. This end of dam ran south to north from west bank but turned beyond two arches in a dog-leg to the right.

Maintenance work on second dam. This view, taken from dog-leg corner on east side, looks south-southeast. Picture was probably taken in 1889. Piers in background face west in same direction as Indian Brave (right center). Upright structure at far end of dam at right not identified. Workmen at left both use single bit axes, and man at right broad axe and peavey. Profile stance of two men was common in early photographs.

This view shows dam built in 1884-85, but date of picture uncertain. Rafter gates were reinforced with splash planks on upstream side of dam. Later, superstructures were eliminated probably when Tainter gates were installed, although splash planks continued to be used on sluice gates and on remaining flood gates not equipped with Tainter gates. At lower left is "bear trap" sluice gate, leaves in horizontal position. Main sluice gate and wanigan shown with arches in center of picture. Note stone cribbing for reinforcement.

Later view of dam shows supersturcture on flood gates removed, but splash planks visible behind gates in center of photo. At lower right is bear trap gate and, at deck level, a winch for raising and lowering water intakes on bear trap.

The "dam house," as it was called, replaced an earlier log shanty for crews at Little Falls. New house also had rooms for transient log drivers and visiting company officials. It was built about 1900 and abandoned after World War I.

View of south end of dam shows boat house as well as main sluice gates at right. Arched entrance to dam house is visible at extreme left. Picture, probably taken on a Sunday, shows men and boys in white shirts hovering over gate to watch wanigan run dam, although at instant camera clicked, wanigan was submerged. Also visible above sluice gate is A-frame used for withdrawing splash planks.

Photo taken in 1913 by Charles Lewis after dam had been abandoned shows Tainter gates in foreground raised. On dog-leg in distance, some gates have splash planks and at least ten gates have splash removed to allow free flow of river. Indian Brave stands at left and "Barney town" in distance at right. At lower right is bear trap gate with winch.

"hurled 400,000,000 monarchs of the forest with irresistible force against bridges, dams, houses and mills, crushing the strongest structures into fragments as if they were built of straw."

In addition to the big sawmill at Chippewa Falls which was seriously damaged, 150 homes were destroyed in Eau Claire, and damage would have been greater, *Harper's* relates, "had not a 'jam' occurred at a spot called the Dells about a mile above the city, and arrested the progress of an enormous mass of logs estimated at 250,000,000 feet." However, the same source said, the gas works and machinery for the dynamo-electric plant was destroyed, and only one bridge out of twenty was left on the Chippewa River. The railroad line from Eau Claire to Wabasha was nearly obliterated.

But the *Harper's* reporter said "the sang-froid of the inhabitants would amaze Eastern people. Already they were at work righting things with a hearty will . . ."

All through the autumn of 1884 and into the winter and spring of 1885, work on the new dam went forward, and there was a constant movement of horse teams and wagons or sleighs hauling rock to the dam site. The rock outcropping in the middle of the river about a kilometer downstream from the dam was nearly blasted to pieces. With the wind sweeping down the ice-covered river from the north, this must have been the coldest spot in Wisconsin for the men who had to chisel holes into the rock formation to insert blasting powder. An embankment of some kind connected the shore line with the rock outcropping to the west side, and the road to the dam followed the west bank north to the dam.

The crews rested on Christmas day, but the blacksmith used the day off to move into his new shop, and two teams came up from Chippewa Falls with supplies. A rare photograph of the new dam under construction held by Henry Plagge (Holcombe), shows two iron tracks less than a meter apart running on raised cross-pieces along the deck of the dam from the west end toward the center. The iron track was used for a cart to bring the rock and other supplies out from shore to the point of construction.

In mid-January Frederick Weyerhaeuser drove up from Chippewa Falls, nearly fifty kilometers (30 miles) in a cutter or sleigh to view the superhuman effort that was being made to have the dam ready for the first log drive less than three months away. He was accompanied by William Irvine, secretary for Chippewa Lumber & Boom, or, as the lumberjacks probably called him, "the push."

In his remarks, the keeper of the daily journal notes that Weyerhaeuser seemed pleased "with all the works and made calculations on sending 40 thousand feet of plank for the dam." Since the dam had no concrete abutments, it was buttressed by huge, box-like cribs of planking and timbers into which rock and gravel were dumped. The planking Weyerhaeuser was ordering could also have been for the splash planks, or for aprons below the sluice gate and wanigan gate.

Weyerhaeuser returned to the dam a second time to make an inspection on June 25, 1885. The keeper of the journal makes no comment except to say that he was accompanied this time by McLeod (Kenneth B.) who, from other references to him, was probably Irvine's right hand man.

In late March, 1885, the dam was nearing completion, and the construction crew, which at times numbered 175 men, was being reduced. On March 23rd this entry appears in the daily journal: "Finished clearing logs off the dam and 17 men sent down." The expression "sent down" means that the men had been released.

The new dam had splash planks in back of the sluice gates as well as the flood gates, and some time after 1890, probably fourteen Tainter gates replaced an equal number of splash plank gates, seven on either side of the main sluice gate. The Tainter gate, invented by Jeremiah Burnham Tainter of Menomonie in 1880, was convex in shape on the upstream side and was supported on two rockers. (See accompanying drawings by Randall E. Rohe). The Tainter gates in photographs appearing in this chapter appear to be made entirely of planks and timbers; the modern Tainter gate, now used around the world, is made of steel and concrete and controlled by various mechanical lifts.

View of Little Falls dam after snowfall. Dam house at far left has been built but gates in dam not yet replaced with Tainter gates. Up to 1906 dam was used as roadway across river.

In addition to the main sluice gate, the wanigan gate, and a narrow gate for the batteaux to pass through, the new dam at Little Falls had two "bear trap" sluice gates, neither one of which was used for sluicing logs. The main bear trap was built at the extreme south end, that is, up against the west embankment, the other on the east end, probably smaller than the other. The bear trap had two large leaves on hinges attached to an apron underneath, and by manipulating water intake through a passageway underneath the apron, the two leaves could be raised to an inverted "V" to catch a head of water, and, when the water was to be released with other gates to flood the river below, the two leaves folded down. No one seems able to explain why this special gate was needed at Little Falls in view of the fact that it was not used for sluicing logs. However, the bear trap could release more water in less time than any of the other gates and perhaps it was for this reason, the fear of another flood like the one in 1884, that the builders installed the two bear trap gates.

Thus, by 1900 the new dam probably included thirty five gates of different descriptions. Beginning from the south end, there was one bear trap gate, seven Tainter gates, one batteau gate, one wanigan gate, one log sluice gate, seven Tainter gates, and then, turning the corner on the dog-leg, eight rafter gates and, finally eight more rafter gates from the small corner of the dog-leg. A postcard photo of about 1912 seems to indicate that the small bear trap gate was located between two of the rafter gates on this end, that is, the extreme east end of the dam.

The living quarters for the crews at the west embankment of the first dam, as well as stables and blacksmith shops, were probably of log construction, but all of these buildings were replaced by frame after the new dam was completed. Photographs taken in the 1890s show a building, referred to as the "boat house" standing at the left of the dam on the south end. This was a two-story building with a basement which was used for storage and repairing of batteaux. From here the batteaux could be launched directly into the river. Upstairs the boat house was used to store supplies consigned to logging camps farther north, and there was also a room for a wanigan, that is, a small store where the men could purchase daily necessities.

On a knoll to the right of the boat house stood a blacksmith shop. To the left of the boathouse, overlooking the south end of the dam but on a higher elevation, stood the new "dam house." A photograph taken allegedly in 1889 shows no house here; the new one was probably built in the mid-1890s. James Jardine, foreman of the crew at the dam in the early 1900s, had an apartment on the first floor, near the center of the building. Kitchen help lived upstairs. There were two rooms on the second floor reserved for visiting officials of the company, and also a larger room with bunks for thirty men of the local crew although there were seldom that many employed here. There was a big day room on the first floor, east end, and on the west end, the dining room. Kitchen and storage facilities were located in a lean-to off the back, and driving crews passing through stopped to eat here, standing at tables outside.

The crew at the dam was self-sufficient as far as possible. In addition to the blacksmith shop, there was a shed for washing clothes, a chicken coop, hay barn, horse barn, a lean-to off the horse barn for two cows, a pig house, an ice house, a smoke house (mostly for the catfish and sturgeon) woodshed and a root cellar which was dug into the ground and covered with sod. This was used to store vegetables and meat in winter, but in summer meat was kept in a big ice box. Ice was cut from the river and preserved in the ice house. And

Henry A. Plagge, Holcombe pioneer, drives across steel bridge built over Chippewa River at Holcombe in 1906. Plagge says picture was taken some time between completion of bridge and "when I got my first Ford in 1915."

Hydro electric dam at Holcombe built in 1948 for Northern States Power Company of Wisconsin. Aerial view looks north-northeast across bridge on County Trunk M. Powerhouse holds three generators, each rated at 11,250 kilowatts. Flowage created by dam covers 4,300 acres and contains twenty-one islands and sixty-one miles of shore line. Little Falls logging dam stood somewhere to left center of picture. Unlike Little Falls dam which had to have watchman year around, hydro electric plant at Holcombe is controlled from Lake Wissota hydro facility.

finally, there were outdoor toilets for both men and women not far from the house, but the bedrooms also had "night jars."

The new dam served the lumbermen for twenty-five years. Chippewa Lumber & Boom, one of its chief users, ran its last log in 1909, but there were still drives by other companies and Zac Jardine says he stood on the dam in the summer of 1911 and watched the last log sluiced through. After that, the logs could no longer be run down the Chippewa owing to power dams which were being built at Cornell and Jim Falls. There was still timber being cut after 1911, and for a few years the logs were driven into the old flowage above the dam where they were sorted at a works near the upper jam piers. After the logs were identified by their end marks, they were shunted into pocket pens and made into small brails which here were merely sack booms held together by circling boom sticks. These booms were towed down the flowage to a landing above the dam on the Omaha Railroad in Holcombe and loaded on cars for shipping out to mills in Eau Claire and elsewhere. A small steamboat about the length of a wanigan with vertical steam engine and paddle wheels, port and starboard, towed the sack booms. About ten to fifteen flat cars a day were loaded with a jammer.

Logs tumbling through the main sluice gate at Little Falls dam. Two ladies and girls, in Sunday dress, and member of crew at left, watch logs coming through dam, while three figures on horizon of waterfall are probably on a boom shoving logs into gate entrance. A-frame on deck of dam was used for retrieving splash planks.

Photograph taken at Gordon (Douglas County) 1903, shows flat cars being loaded with jammer on cross-haul. Pine logs on car scarcely large enough for studdings, but big log on ground has been found for benefit of cameraman. Jammer, often called "woods jammer," or "side jammer" or "horse jammer" is mounted on runners and can be slid about by team of horses. Man in center below skids is probably scaler and holds rule stick to measure board feet of lumber in every log.

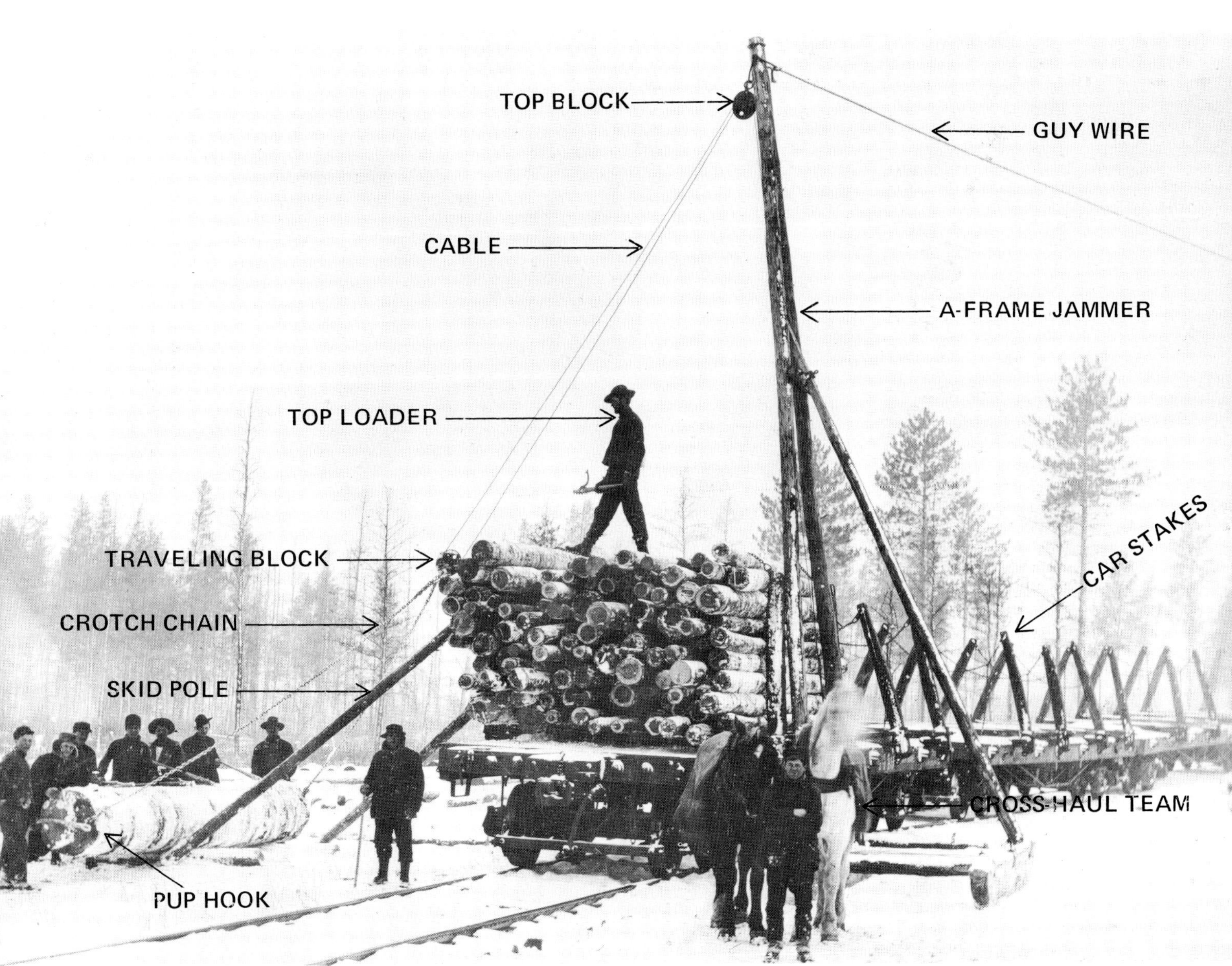

Wanigan, probably largest ever launched on Chippewa River, followed log drives in 1890s all the way to West Newton slough for Mississippi River Logging Company, and was towed back to Eau Claire by steamboat. Crew ate in mess served on board, not ashore. Tom Lucas, head cook, stands in center, arms folded, flanked by second cook. Photo probably taken on Minnesota side.

BEAR TRAP GATE

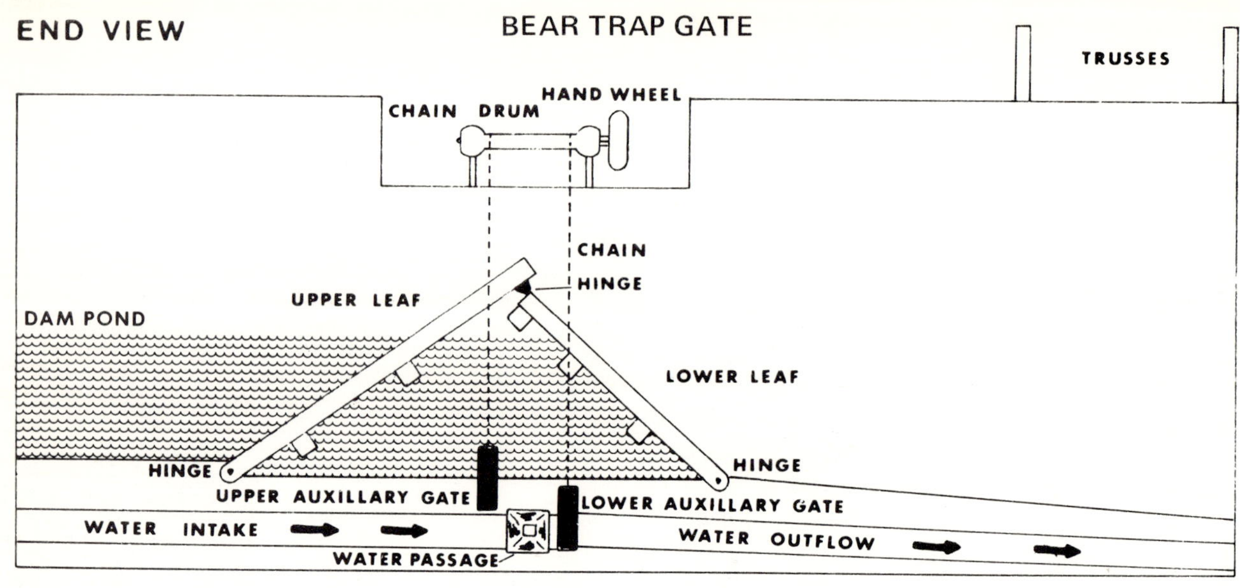

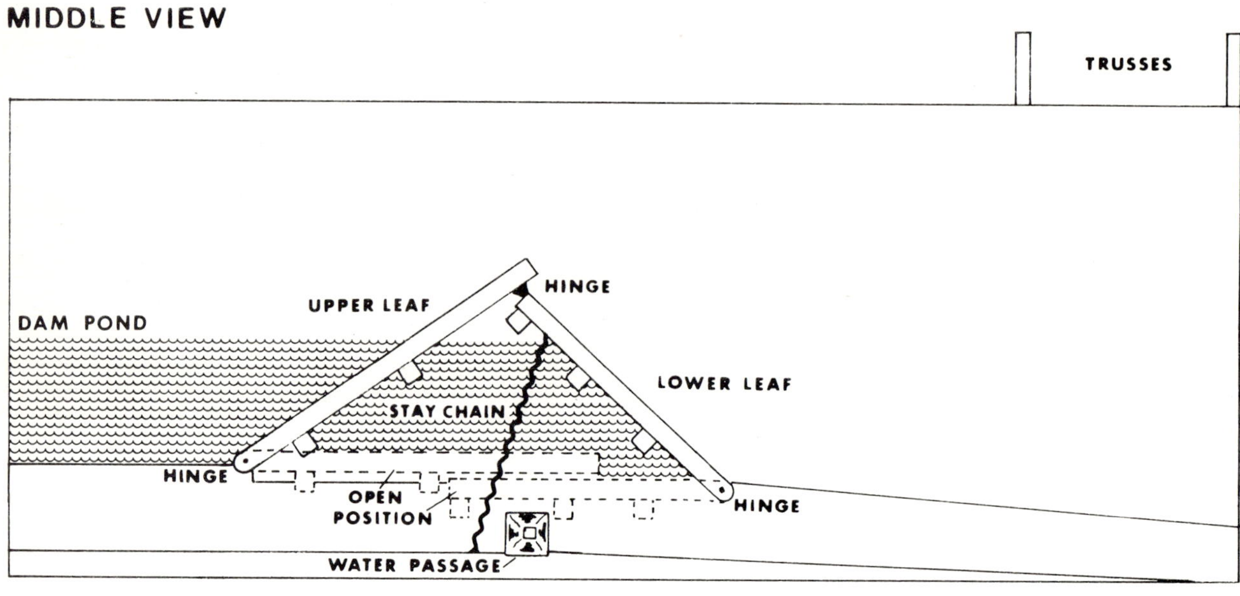

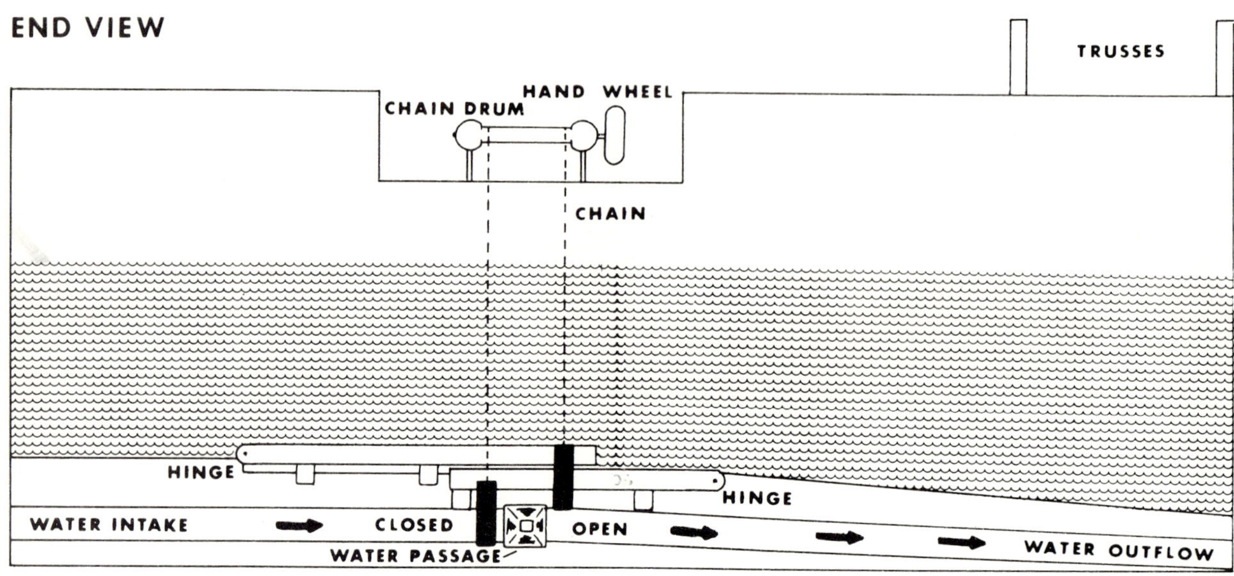

Sketches by Randall E. Rohe and Zac Jardine show different views of bear trap sluice gate. Built with two rectangular leaves secured to bottom of apron of dam by hinges, gate was raised by pressure of water from pond above which was conveyed into duct controlled by small gate to chamber under apron. A second duct, with stopcock, connected chamber with pool below dam. When duct from pond above was opened, lower duct closed, and water filled chamber under two leaves causing downstream leaf to rise by flotation. Bear trap dam was developed in Pennsylvania in 1818 and in 1880s and '90s, was popular on both large and small rivers for log driving. Nevers Dam on St. Croix River, built in 1890, had bear trap gate eighty feet wide.

Top: Bear trap sluice gate at south end of Little Falls dam on west bank. This was special gate which was used not to sluice logs but as a safety waste gate. Picture shows gate at low water. Two leaves which were raised to an inverted V position to raise head of water, are here lying flat on apron. Dark horizontal line running across center of leaf is fixed hinge on lower leaf (see drawings opposite page). Duct underneath leaves carried exhaust water out of gate through exit at left. Photo taken in June 1914 after dam at Little Falls ha l been abandoned, and stump rests in center of sluice. Roof of boat house at upper left. Below: east end of dam in picture taken in June 1914 after dam abandoned. Tainter gates have all been removed. At left center may be remains of small bear trap gate.

Photograph shows flowage for log storage above Little Falls dam, looking west-northwest, since Chippewa River approaches dam in big bend. Booms, barely visible at far left and plainly visible at lower right were used by pole men to walk on while pushing logs into main sluice gate at lower left (not visible here). Booms at right actually abut one of dam gates, while booms at left follow west shore, since sluice gate was closer to west bank of river than to east bank. Pole men in picture are using long pike poles to shove and guide logs into main chute. Two men can be seen atop logs at right and two at extreme left. In high water, logs often floated into rampikes seen at upper right which, for some reason, were never removed. Flowage behind Little Falls dam extended almost to mouth of Flambeau River.

View of bear trap on south end of dam with Tainter gates at right.

This photo probably taken below Jim Falls on Chippewa River about 1903. Seated crossed legged is "Indian Joe," a Chippewa who "could walk under water but couldn't swim." Standing at left of Indian Joe is "Muskrat Joe" DeBault (pants open). Behind him in big straw hat is Billy McIntyre. Holding dish in center is "Spotty" Dan McDonald. Behind McDonald at left (wearing mustache) is Jim Hedrington. In center, right, hands in pockets and smoking pipe is "Red" Paddy McDonald. From right, reading left, stand Scotty McTavish, Archie Smith (arms akimbo) who was river boss, Charlie Browning, Johnny Burke (short man in front of Smith), "Big Pete" Murphy, Ray Smith, a timekeeper, and Billy ("Boo-hoo") Hoyer. Others in picture not identified positively include Charlie Ermantinger, George Craig, Bill Smith, Alfred Smith, and Joe Violet, a Chippewa.

Driving crews passing through Little Falls were fed at stand-up tables behind kitchen of dam house.

Making hay for company cattle at Little Falls camp. Thadeus Loiselle is on load tramping down, and Charles ("Kib") Ecker and Albert Waben are pitching. Horses are "Dan" and "Frank".

Holcombe Methodist Ladies Aid en route to meeting accompanied by local pastor.

Outside dam house a tent has been pitched for children to play. Men are part of the crew attached to Little Falls dam. Left to right, Albert Glenn, Henry Gysbers, Charles Ecker, Arthur Gysbers, Albert Waben, a Chippewa, James Birmingham, Benjamin Fern, James Jardine, foreman, Mrs. Jardine, head cook, Mrs. Bessie Lamb and Nora Lamb, assistant cooks. Seated on ground are Tom, Jim and Zac Jardine, and Clayton Lamb (in white shirt).

Pioneer women of Holcombe in photo taken about 1905. Front, left to right, first unknown; next Lola Tinker, and next in dark dress (white collar), Florence Moore, a church deacon and lay pastor; next unknown, and last, Mrs. George Spafford. In second row, left, Mrs. Caroline Dehler, Mrs. Roy White, Mrs. Omar Adams, Mrs. Phoebe Edminister, Mrs. Orin Adams, and Mrs. James Jardine who holds son Tom, and last unknown. In third row, left, Mrs. A.J. Edminister, Mrs. George Zerbach, Angeline Zerbach, Dell Brooks, and Mrs. Walter Dehler with child. Fourth row, left, first unknown, next Mrs. Thadeus Loiselle, Mrs. Alfred Brunning, next unknown, and last, Edna West. Three children down front not identified.

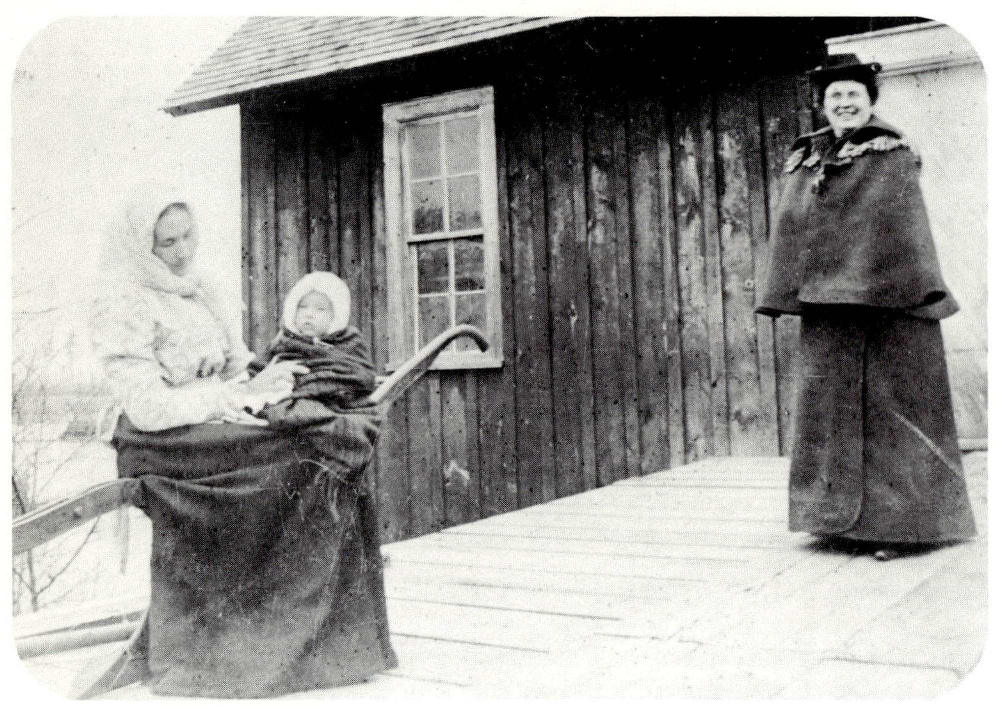

Standing on ramp of boat house at Little Falls dam is Adelaide Irvine, wife of William Irvine, superintendent for Chippewa Lumber & Boom. The Irvines have driven up from Chippewa Falls either by buggy or cutter and, while Mr. Irvine has gone elsewhere on business, Mrs. Irvine visits with Mrs. James Jardine, seated on beam of breaking plow holding son Zac. Picture taken about 1900.

LITTLE FALLS DAM & CAMP 1910

(HOLCOMBE)

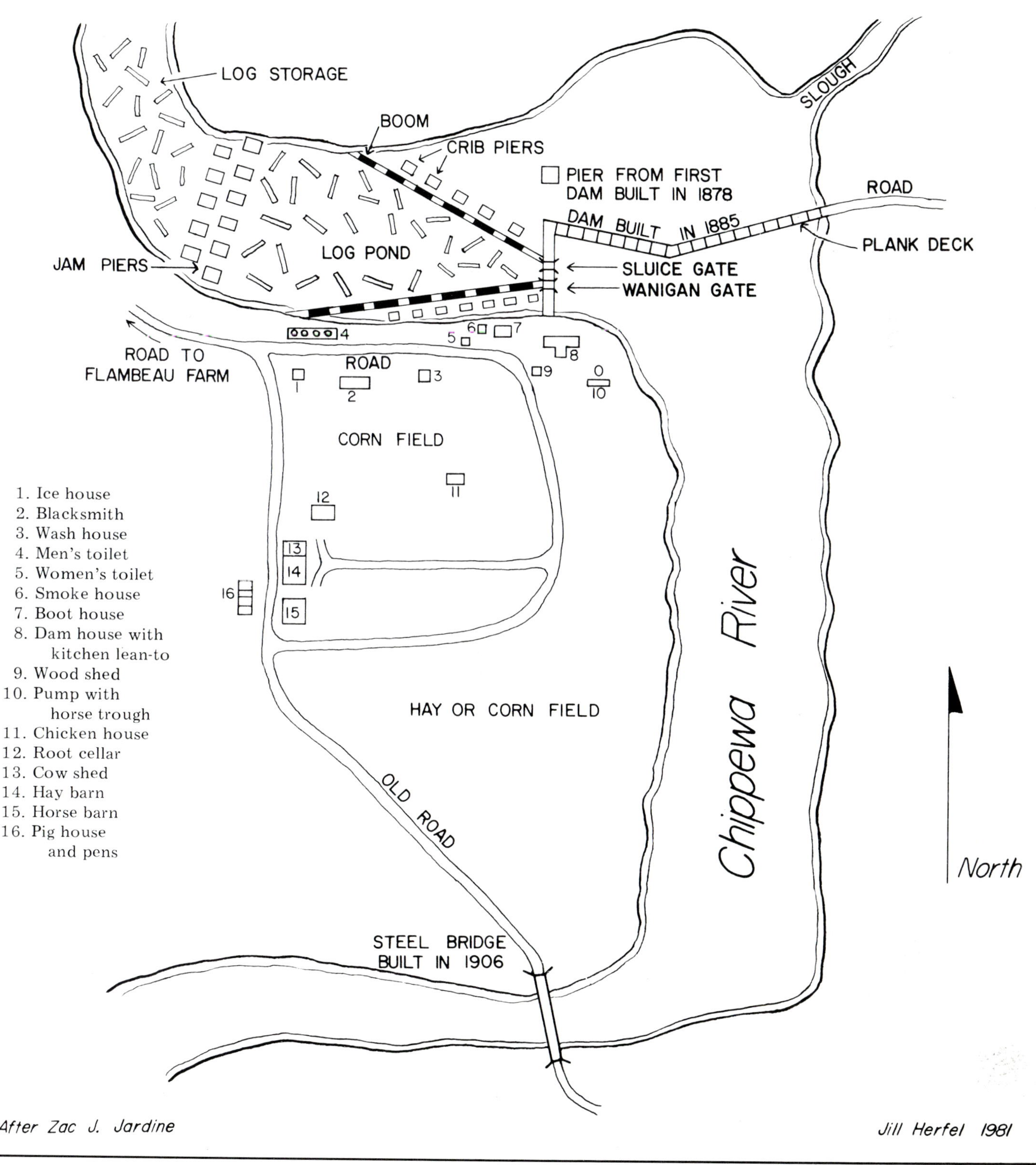

After Zac J. Jardine

Jill Herfel 1981

Journal Of A Dam

At sea a captain of a ship keeps a log. At the Little Falls dam on the Chippewa, the foreman of the dam crew kept a log, or journal, in addition to an hourly account of the water level in the pond above the dam. Like the captain's log, this was not a personal diary. It was a record required by the Beef Slough Company which maintained the dam.

The daily entries in the journal were made in a big ledger-type book (see accompanying photo) and began on June 22, 1882, and ended on August 16, 1890. There is no name in the journal to identify the keeper of the journal, but the entries are made in pen and ink, and, in the course of eight years, the penmanship changes three or four times. The late Francis Gannon of rural Cadott discovered this historic document in a box crammed with old books and papers purchased at an auction. Mrs. Leslie Jones (Holcombe) transcribed it for the Chippewa County Historical Society, and the society published it as a booklet.

It would seem, from the context of the remarks made in the journal, that the foreman who made the entries was also responsible for directing the crew to open or close the gates when the logs were ready for sluicing, and also to keep the gates in repair. Many times he had to start his crew sluicing at midnight, or at three in the morning, and on several days the men worked sixteen hours without a break.

The journal is included at the end of this chapter and the introductory remarks that follow are made to clarify some of the nomenclature and to provide background information. For example, the verb "to sluice" is not found in modern dictionaries of the English language. In the logging era, it meant running logs through a dam, and the gate where the logs passed through the dam was called the "sluice gate." There was usually only one sluice gate in any dam; the other gates, whether two or three, such as found on the smaller rivers, or multiple gates such as found on the Chippewa and St. Croix, were flood or waste gates, that is, gates which were closed from time to time to build up a "head of water," or opened to release a head in order to begin sluicing.

On July 15, 1882, Louis Musler came down to Little Falls from Jump River and "left kit in my charge." Two days later the journal says this: "dried out Jump kit and packed it in new warehouse." In this instance a "kit" was a bag or box which contained the tent and blankets used by a crew on a log drive.

On July 22, 1882, this entry appears: "Cook came up for a mealer. Sent him up all right." Apparently a jobless cook had come to the camp where the crew lived on the west end of the dam and asked for a handout. It seems he got the "bum's rush," although in the 1890s and early 1900s it was seldom that anyone was turned away hungry from a camp in the woods.

Several place names are mentioned. Joco (also as Jocco) pronounced Joe-ko, was an island probably lying north of the jam piers and it was named after a Chippewa Indian who worked for many years on the drive. The island is now submerged.

Reference is made more than once to "Surveyor's Point." This probably lay on the west bank of the flowage opposite the mouth of big Jump River.

"Elum" [Elm] Bottoms is mentioned. There was a liquor tavern in the 1930s called Elm Bottoms below Radisson on the Couderay River, but the name may have been borrowed from a locality farther down the Chippewa River, probably close to Windfall Lake.

"Devil's Nest" which appears in the journal was probably located in Section 1 of Holcombe

Crib piers on upstream side of Little Falls dam served as anchors for booms running from right to left which created channel for logs approaching main sluice gate (barely visible) in left center. Black diagonals at left are Tainter flood gates. In center distances stand boat house and farther up slope at left is gate leading to dam house for crew.

township, or about fifteen kilometers up the big Jump River. But references to "Jump Islands," "Devil's Hole" and "Sugar Bush Dam" appear to be lost in time. "Fisher Dam" is mentioned and this lay on Fisher River (NE-NW Sec. 34) or about three kilometers southeast of Holcombe.

There is mention in the journal to "LeBeouf." This was probably Joseph LeBeouf, a Frenchman who had a logging camp a short distance north of what became the village of Winter in Sawyer County. LeBeouf's was also a stopping place for sheriff's deputies en route from Hayward to capture the defiant John F. Dietz at Cameron Dam on Thornapple River.

In several entries, the keeper of the journal refers to men about to *paal* [pole]. These were men with log pike poles who walked along shore to push logs away from the banks into the current of the river, or they were pushing logs into sorting gaps, or into a sluice gate.

Reference is made to a "running drive." This was a drive of logs that went downstream with no thought taken for strays. The idea was to get to the sawmill as fast as possible with as many logs as possible.

"To catch a head," a term mentioned elsewhere, refers to closing the flood gates on a dam and raising the level of the water in the pond before beginning to sluice logs.

Mention is made of a "pole tax," a misspelling for "poll tax." Before 1910 every family living along a town road had to contribute one man one day a year to highway maintenance. If no member of a family could find the time to work, he could hire someone to work in his place, or he could pay the penalty which was about $1.50. Without payment or work, he legally could not vote, but there does not appear to be many instances where anyone was disqualified on the grounds of not having worked out his poll tax.

There is constant reference in the journal to a "man to town with dispatch." There were no telephones and a courier had to carry messages from the foreman at the dam to Chippewa Falls for instructions. At Chippewa Falls there was limited storage facilities in the booms at Paint Creek Rapids and there had to be coordination between the mill foreman and the foreman at Little Falls to keep the mill supplied with logs at all times, not too many, not too few.

Kenneth B. McLeod, river boss for Chippewa Lumber & Boom, arrived at the dam from Chippewa Falls on April 27, 1883, with orders to begin sluicing at night. This seems extraordinary, but it was probably done to avoid high winds during the day. Wind from the south or east pushed the logs backward and made them hard to handle. (McLeod was born in Inverneshire, Scotland, in 1849, and came to Chippewa Falls in 1869 from the Au Sable country in Michigan. He

1884

July

Remarks

" 1	Stoped Sluicing and prepared for Flood
" 2	Flooded Waste flood
" "	Part of crew went up on Drive at Flambeau to put in Rear on Flambeau River
" 3	Recived Orders to cut down the wages
" 4	Shut down for Fresh Head
" 6	Rear all down to mouth of Flambeau Pulled jam at Jump Islands
" 8	Moved Wanigans down and Pulled Jam over Rapids at Flambeau Farm
" 12	crew Puting in Rear Watter very Low Dam all Shut Down For Fresh Head
" 14	Flooding 12 P.m.
" 15	Ran Boats Down to Wanigan and Prepared to suspend Drive Robe
" 16	Water very Low Paid all the crew at Flambeau Farm
" 17	Went down to Town

Mrs. Joyce Gannon holds ledger journal kept at Little Falls dam.

began working for Weyerhaeuser in 1871.)

On July 18, 1883, this entry appears: "men went to work on dry roll." This refers to a special crew, not the rear crew although it also worked in the rear of a drive. After the rear crew had sacked the logs stranded in the sloughs and on the sandbars into the river, there were often logs farther back from the embankments which the rear crew could not handle. A special crew was then sent to "dry roll" with horses or oxen. A short spike with an eye in the head was driven into each end of a log and chains went from the eye to the single tree behind each horse (or oxen). When the two horses pulled on the chains, the log rolled forward like a wheel. (See accompanying photo). And because these logs lay on higher land after a flood receded, the land was, presumably, dry, and hence the expression "dry roll."

On August 31, 1883, a journal entry advises that the crew at the dam had "opened 6 gates to flood Whidden over Brunies Falls." Brunet (pronounced Brunee) Falls was a constant hazard on the Chippewa River. It was located about five kilometers downstream from the Little Falls dam. John Whidden was a boss on the drive, and with a little extra water from the dam at Little Falls, he got his logs over "Brunies" Falls. William Hoyer remembered Whidden whom he referred to as a "bull of the woods."

Fred Pitsch (or as Pitch) is mentioned several times in the journal mainly because he was driving the "tote team," that is, a team of horses which hauled supplies into camp, usually from Chippewa Falls. On October 29, 1883, Pitsch arrived in camp with a load of coal, almost surely for the camp blacksmith, not the cook.

On February 10, 1885, this entry appears: "Had a visit from Mr. Stafford, ticket agent for the Eau Claire or American Hospital Association." This was a new service for the lumberjacks. Tickets, usually costing $5 to $10 a year, were sold to the jacks as insurance, and the holder was entitled to medical care in a hospital designated on the ticket. There was little illness in the woods camps; most of the patients who reached the hospital were there as a result of accidents, and these could be brutal.

The names of at least a couple dozen men or bosses at the Little Falls dam are recorded for the years 1882 to 1890, but there are two which will probably never be identified. On May 11, 1886, this entry: "Porkey Hog came up to drive Alder Creek," and, on May 12, 1889, "Squint Eyed Billie got his drive out of Alder Creek."

Although the keeper of the journal always referred to "boats" around the dam, the more common designation for boats on the Chippewa in the late 19th Century was "batteaux," plural for batteau.

In mid-May a big jam occurred on the holding pond above the Little Falls dam, and the crew, according to an entry in the journal, was "picking" at this for days, and then another entry has this to say: "no haul on the jam yet." The use of the word "haul" in this context is rare, but it means that the crew had not managed to break the jam. A few days later, one of the crews "got a big haul on the jam at Flambeau too."

The expression to "pick at a jam," as noted in the above, is used elsewhere in the journal and it suggests that a big log jam was not broken by getting at a "key" log. A jam extending clear across a river had to be picked at, one log at a time, usually with a long cable or line running to a team of horses ashore. William Hoyer, in personal conversation with the author, refers to a key log in reference to a "center," which is correct because a center was not a jam extending across the river, only to logs caught around a stone or sandbar in midstream.

On April 14, 1886, the journal refers to splash planks which were taken off "one roll on other end of the dam." Three days later, splash planks were

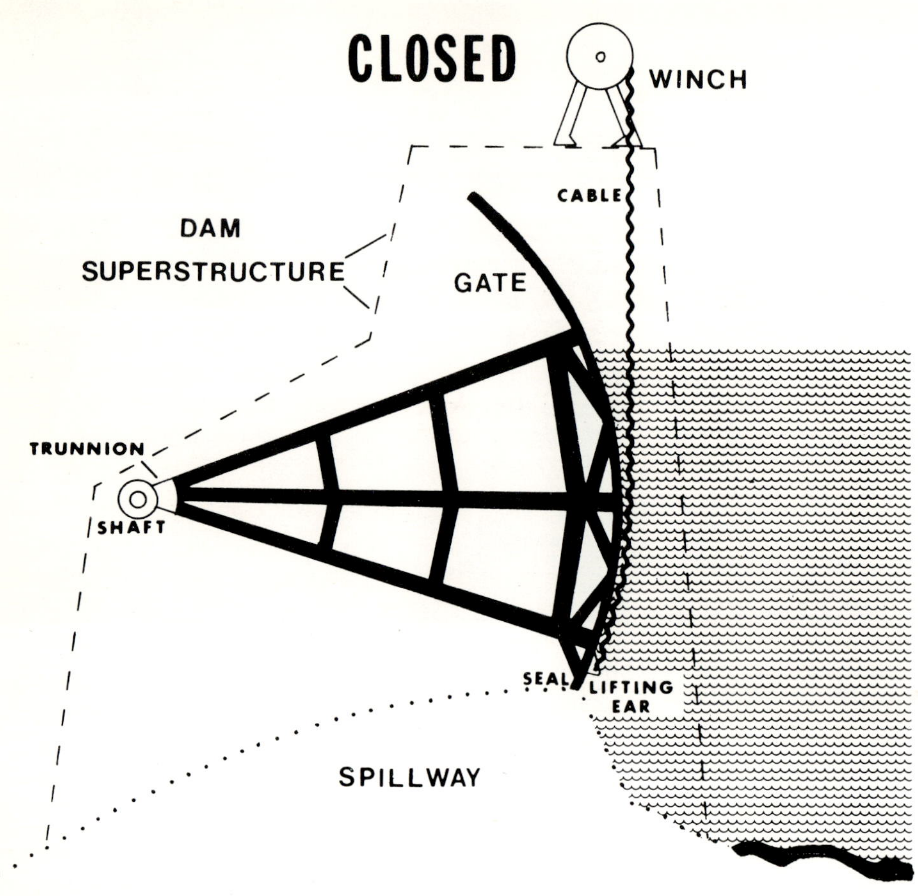

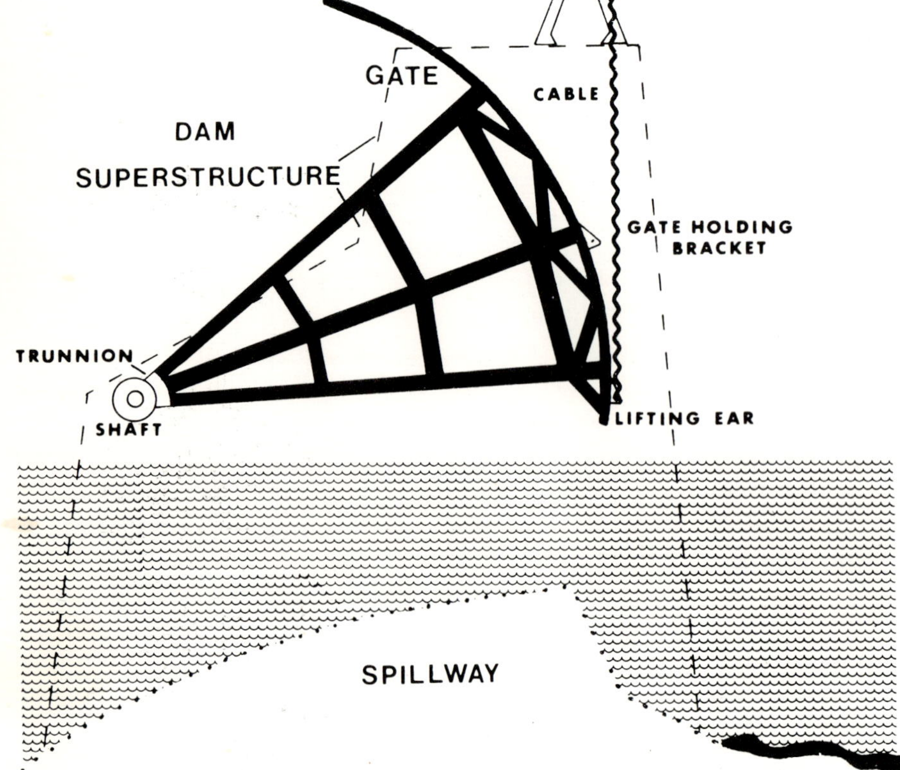

Drawings at left were made by Randall E. Rohe to show Tainter gate for logging dam invented in 1880 by Jeremiah Burnham Tainter of Menomonie, Wisconsin. Early gates were made entirely of square timbers and planks and were raised or lowered manually by winch from deck of dam. Gate operated on rocker principle with convex side facing upstream. Modern Tainter gates are constructed of steel and concrete and are used on rivers around the world.

put on the "small roll." Thus, it seems there were two gates referred to as "rolls." (Splash planks were also called "flash boards" on some rivers but not on the Chippewa.)

In 1918, F.A. Pott, a surveyor for Wisconsin-Minnesota Light & Power Company, established bench marks on the Little Falls dam, now abandoned, and in his report he refers to the water crest on a "bear trap dam" (gate), and shortly later, to the crest of "second bear trap dam." This bit of vital information suggests that the keeper of the daily journal at the dam referred to the bear trap sluice gates as "rolls," not as bear traps.

The bear trap sluice gate for sluicing logs on driving rivers was developed in 1818 by Josiah White and Erskine Hazard, managers of the Lehigh Navigation Company in Pennsylvania. Onlookers who were watching the two engineers work on this strange looking device wondered what it was, and, to satisfy their curiosity, the engineers told them they were working on a new type of bear trap, and the name stuck.

The bear trap gate consists of two rectangular leaves of a length equal to the width of the opening in which they are installed. Each of the leaves had an axle or hinge which enabled it to move or slide up and down. Some of the smaller bear traps had a roller on the lower leaf, but whether roller, axle or hinge, all were installed to enable the leaves to move or slide up and down. When the gate was down, the upstream leaf overlapped the downstream leaf.

The bear trap gate is actually raised and held in place by water pressure from the pond above, the water being channeled under the apron of the dam through a duct which carried the water to an opening under the leaves, and another duct which could be opened for the water to run out. When the water rose under the two leaves, they raised, the pressure under the leaves being equal to the pressure in the pond above. To lower the two leaves, the downstream duct was opened and the water escaped. With the leaves down, an enormous amount of water rushed through the gate.

Owing to the failure of a big bear trap dam built in France, bad publicity followed its original development and it was not until many years later that it was revised, especially for logging camps in Ohio and in the Lake States. One engineer, Ashbel Welsh, president of the American Society of Civil Engineers, told a convention audience in 1882 that the bear trap was an excellent dam and "possibly the cheapest, for letting the water rapidly out of a reservoir for scouring purposes." By "scouring" he probably meant flooding or releasing a "head" of water.

Word of this may have reached the engineers of the Little Falls dam who, in 1885, installed two bear traps. And when the Nevers Dam was built on the St. Croix in 1889-90, it had only one sluice gate and this was a bear trap eighty feet wide. (Wider bear trap gates were built.) But unlike the Nevers Dam, Zac Jardine is convinced that the bear trap next to the boat house at Little Falls was never used for sluicing logs, and the daily journal kept at the dam in the 1880s seems to confirm this. Here it was strictly an auxiliary flood gate.

The weather was a source of almost daily interest in the journal. On November 15, 1888, the first snow fell and on the day after it was "colder than Hell and damnation."

In the last entry for 1888, which fell on November 22nd, the journal says this was "splended weather for working on the apron (and) Smith gone down the road to hire another team." This was probably Archie Smith, a river boss for many years for Chippewa Lumber & Boom. He was still on the river when William Hoyer began working for him around 1900, and he bossed the very last drive for the company in 1909. Hoyer remembered him. Said he: "I know why he never got anywhere, because all he used to do is sit and whittle about a mile away from the river. He didn't go with the crew at all. He didn't smoke or anything. Just sit all day and whittle . . ." But Smith was on the river before Hoyer was born!

In the following transcription of the daily journal, no attempt is made to correct the spellings or language as it appears to be written. The men who kept the journal were literate only to a point, nevertheless, most of them wrote with a clear and legible hand, and, considering that they were using steel pens which had to be dipped in ink, the overall result seems rather remarkable.

RECORD FOR JUNE 1882

June 22, Heisted gates. Let flood of[f] 12 o'clock P.M.
June 23, Let water run all day. Hevy rains up north. Wind N.W. One in. on level hear.
June 24, Clear & warm. 6 o'clock P.M. shut 5 gates. Loggs coming in fast.
June 25, Water rising 4 ft. below dam. 12 ft. on dam 6 o'clock P.M. Closed 2 gates.
June 26, Water rising 4 ft. 6 in. below. Water in pond 15 ft. Opened ___? 8 o'clock. Wind S.W.
June 27, Commenced sluicing 4 o'clock A.M. Sluiced 16 hours. Water in dam 15 ft. 6 in. Water below 5 ft. 5 gates up. Fred Pitch arrived with load shingles.
June 28, Water falling 4 ft. below. In dam 15 ft. 6 in. Rain fall ½ in. 4 gates up.
June 29, Flood 12 o'clock P.M.
June 30, Water run all day.

JULY 1882

July 2, Rain fall A.M. Wind S.E.
July 3, Water rising in river above.
July 4, Clear & pleasant. Water rising.
July 6, Commenced driving on Jump [River]. Brook [Broke] jam.
July 7, Rain fall 10 P.M. Half in. Pitch with load of shingles.
July 8, Loggs running through pond.
July 9, Rain fall ¾ in. Wind N.W. Six men from Whidden (?) to work on Jump.
July 10, All well. 13 men working on Jump jam.
July 11, Heavy wind N.W. look like rain. Maid flood.
July 12, Pitch with load of suplies. 6 P.M. Look like rain. Shut 12 gates 11 P.M.
July 13, All well. High winds.
July 14, Had flood from Jump. Logs running in.
July 15, Seven million ft. in pond. Louis Muslar came in from Jump with crew. Left kit in my charge. Took old boat. Started down river. Rain fell ¼ in. S.E. Rain fell two & half hrs.
July 16, Present high wind with light shower.
July 17, Dryed out Jump kit & packed it in new warehouse. Rain fell ¼ in. Wind S.W.
July 18, 13 men from Chippewa Falls to paal [pole] in pond. 1 o'clock commenced to fit up wanigan. 6 P.M. J.S. Barnett with six men and Louis Mzsular to tak charge of Jump and pond rear. Louis Mishlar commenced working on pond. 23 man total. Dam team at work filling up east wing. Water rising verry slow. Expect water from upper dams.
July 19, Water not come yet. Expect every hour. Pleasant wether. Wind N.W. by N.
July 20, 6 P.M. Two teams with suplies for pond drive.
July 21, Light shower. Sent boat load with suplies to pond drivers.
July 22, Cook came up for mealer. Sent him up all right.
July 23, All well.
July 24, Pitch with load suplies.
July 25, 2 o'clock A.M. left for Chippewa Falls. P.M. Light showers. Wind N.W. Put man on Fisher dam to watch it.
July 26, Rain fall ¼ in. Rain 2½ hours. Wind N.W. 2 o'clock. Pitch with load supplies.
July 28, McLeod with order to shut down.
July 29, McLeod for Chippewa Falls. To [two] teams with lumber from Chippewa Falls. Team from Stiles [stage line].
July 30, Comenced sluicing 2 o'clock A.M. Poor work. Headwind. Rain fall ½ in. Wind S.E.
July 31, Rain fall 1½ in. Wind S.E. Pitch with load suplies.

AUGUST 1882

Aug. 3, Pitch with load of suplies. Mishler quit driving. Went down with crew. Rain fall ¼ in. Wind S.E. Thirty million ft. logs in pond. Dispatch from Chippewa Falls want logs & water.
Aug. 5, Three men started up Jump to make hay.
Aug. 6, All well. Wind S.E. Blowing quite frish.
Aug. 7, Comenced sluicing. head wind. Poor work.
Aug. 8, Louis Mishler pased Little Falls on his trip up Jump River.
Aug. 9, All well. Water falling fast.
Aug. 10, Had to stop sluicing on account of low water. Shut up small roll. Had to rise head. Expect to flood Sunday night.
Aug. 11, Rain. Wind S.W. 2 hours ½ in. Finished making hay on Jump.
Aug. 13, Rain. Wind S.W. 2 hours ¼ in.
Aug. 14, Shut 24 gates.
Aug. 15, Rain fall ¼ in. Wind S.E. McLeod dispatch changed date of flood.
Aug. 16, Louis Mishler stoped on his way down river. Crew started to work on road.
Aug. 18, Pitch with gering wheels for dam.

SEPTEMBER 1882

Sept. 11, Rain fall ¼ in. Wind N.W.
Sept. 12, Pitch with load of lumber.
Sept. 13, Dam team to Chipewa Falls for load of feed.
Sept. 15, Pitch with load iron. Water from upper dam at 2 A.M. raised pond 2 ft. 6 in. Logs coming in fast.
Sept. 16, Rain. Wind S.W. Fall ¼ in.
Sept. 18, Rain fall lasting 2 hours ¼ in.
Sept. 19, Cold. Rain. N.E. Pitch with load lumber.
Sept. 20, Light showers. Wind S.E.
Sept. 22, McLeod with dispatch wanting logs and water.
Sept. 25, Smith and Henry with crew from Flambeau drive.
Sept. 26, Left for Chip. Falls by road. Left kit at Little Falls dam.

OCTOBER 1882

Oct. 2, Rain fall ¼ in. lasting three hours. Wind S.W.
Oct. 3, Rain fall 1¾ in. Lasting three hours. Wind S.W.
Oct. 4, Rain fall 2 in. lasting six hours. Wind S.E.
Oct. 5, McLeod with orders shut down gates. Rain fall ¼ in. lasting three hours Wind S.W.
Oct. 6, Crew came in from road. Commenced sluicing. Pitch with load of lumber. Brought down boats from Flambeau 14 in all.
Oct. 8, Rain fall 1½ in. lasting six hours.
Oct. 9, Drawed water down. Opened big roal.
Oct. 10, Shut down for fresh head.
Oct. 11, Louis Mischler with 36 mean to drive. Got kit at Little Falls including boats. Run wanigan over big roal. Full head on.
Oct. 13, Dam team at work halling kit over Brunnets falls.
Oct. 14, Too men at work at Fisher dam.
Oct. 15, Rain fall lasting too hours ¼ in.
Oct. 18, McLeod orders to flood.
Oct. 19, Preparing for flood. Logs running in fast. Light showers lasting one hour.
Oct. 20, Pitch with load lumber. Dam team with load suplies.
Oct. 21, Light showers.
Oct. 22, Rain fall lasting four hours ½ in.
Oct. 23, Drawed pond down to fix dam.
Oct. 24, Comenced to repair gates.
Oct. 23, Pitch with load of coal.
Oct. 30, All right.
Oct. 31, 3 men sent up to work on dam.

This is one of last years Little Falls dam was operational. Tree has grown in front of bear trap gate at right. Arrows point to splash planks protruding above deck of dam. Indian Brave stands at left.

APRIL 1883

April 15, Ice comenced to run. Water rising. Snow went off in 3 days. Heavy ice jams on Jump Island. Big jam on Jump.

April 18, McLeod with men to work on Jump Island.

April 19, 9 men to drive on Divls nest. Caligan with crew to brake jam on Chippewa.

April 20, Water falling in river. Rain and snow fall, lasting 9 hours. Wind N.E.

April 24, McLeod with orders.

April 25, Shut small roal.

April 27, McLeod with orders to sluice nights.

April 30, Little Falls team with supplies. Sent man to Jim's falls. Falls all right.

MAY 1883

May 1, Sent team to Jim's falls. Rain fall 5 hours ½ in. Wind S.E.

May 2, Jump Island jam all through.

May 3, Caligan left for Chippewa Falls.

May 4, Crew from Devil's Nest to rear on Chippewa.

May 6, McLeod from Silver Crick with crew. Left kit at Little Falls without receipt.

May 7, Drawed watter in pond to 9 ft.

May 8, Shut down for head. Logs coming in fast. Watter rising. Davis crew driving past Little Falls with boats and kit.

May 9, Rain fall lasting 9 hours fell 1 in. Little Falls the crew 14 men went on Jump Island to take in rear.

May 10, Ran the jam through from Jump Island.

May 11, K.D. McLeod orders to hold logs. Shut the booms and jam piers.

May 13, Logs comming in fast.

May 14, Jam of seventy million in pond near at Surveyor [Point]. Evening river rose 11 ft. by flood from upper fall of rain lasting 9 hr. fall ½ in.

May 15, Little Falls team left for town. Opened booms and sluiced into dam.

May 16, Team from town load of supplies.

May 17, Rainfall lasting 12 hr. 1 in fell.

May 18, Rainfall lasting 18 hr. 1½ in. fall.

May 19, Rear of jam mouth Dear Tail [creek].

May 20, Crew stoped working on jam.

May 21, Little Falls crew to work clearing pond.

May 23, Little Falls team up load lumber.

May 24, Crew went to work on rear. Rain fall lasting 6 hr. 5 in. fall.

May 25, Little Falls crew to town. J. Peral went over to Fisher locate a dam.

May 26, Heave shower. Rain. Wind S.W.

May 27, J. Peral left orders to sluice on the 27th. Ingrams team came up from Eau Claire. McCafoms kit.

May 28, 6 men left here to work on rear.

May 29, Shut booms and stoped sluicing, and Remaude crew went to Brunies Falls to take [care] of jam. J. Peral returned from Fisher and dicied not to build a dam. Little Falls team started to town.

May 20, Drawed pond 8 in. to give water at Brunies falls where crew to work on jam. Brunies Fall all cleared. Crew returned to Jump Island.

May 31, Jam pulled on Flambeau Rapids 1 mile. Sent 1 man to Jim's Falls to see if river was clear. Commenced sluicing. 18 hours. Team left Little Falls for town.

JUNE 1883

June 1, K.D. McLeod with orders for flood. High S.E. wind. No sluicing.

June 2, Rain fall lasting 14 hr. ¾ in.

June 3, Drawed water in pond down to 10 ft. That is a waste flood. Run wanigans over Little Falls. Tied up at Brunies Falls by order K.D. McLeod.

June 4, Rain fall lasting 5 hr. ¼ in.

June 5, Little Falls team with load lumber. Light showers. Dam crew work on road.

June 6, Took crew from road to work Jump Islands.

June 7, Opened booms, Sluced 6 hr.

June 8, Commenced sluicing and continued to 16 hrs.

June 9, Part of the crew working at Jump Islands.

June 10, Rain fall lasting 5 hrs. ½ inch.

June 11, Run jam through jam piers and crew went back to Jump Islands.

June 12, Received orders to hold logs.

June 13, Held logs by orders.

June 14, Little Falls crew to work on road.

June 15, Rain fall lasting 5 hr. ½ in. fell. Little Falls team whent to town.

June 16, Heave showers lasting 1 hr.

June 17, Little Falls team with load supplies. Rain fall lasting 9 hr. ¾ in. fell.

June 18, Rain fall lasting 4 hrs. ¼ in. fell.

June 19, Stop work on pond and prepared for sluicing the 20th.

June 20, Commenced sluicing and Slu[iced] 16 hrs.

June 21, Water falling fast in river.

At right, wanigan slides through sluiceway at Little Falls dam.

June 22, Sent part of crew to Jump.

June 24, Sent two men to Flambeau for wanigans.

June 25, Water falling fast in river. Strong north east wind and could not sluice. Took crew and sluic at night.

June 27, Jam all below dam. Whidden wanigan down to the dam. Took driving kit and put in warehouse also 18 boats tied at dam.

June 28, McLeod with orders to take in rear from Flambeau Farm.

June 29, Put stuff on wanigans and L. Lyons left here with canoe.

June 30, To put rear in at Flambeau.

JULY 1883

July 1, McLeod orders for a flood. Wind S.E.

July 2, Rain fall lasting 14 hrs.

July 3, Heavey showers lasting 1 hr.

July 4, Water rising in river. Light showers.

July 5, Shut up small rool.

July 6, Heavay fall rain lasting 16 hrs. 2 in fell. Little Falls team with load plank. Opend big rool. Water raising fast.

July 7, Two teams came up to work on dry roll.

July 9, Little Falls team went to town. Opened booms & jam piers and sluiced 18½ [hours]. Team up from town arrd. and five men to work on rear.

July 10, Received word from K.D. McLeod to hold logs and prepare for flood.

July 11, Little Falls team with with load plank. Light showers.

July 12, L. Lyons team went up to work on rear.

July 14, Rain fall lasting 14 hrs. 1 in. fell.

July 15, All right and very mild.

July 16, Grand strick [strike] for more pay. Luk Lyons goes for a new crew. Rain fall lasting 8 hrs. ½ in. fell.

July 17, Crew from town to work dry rool. Wages $2.00 per day. Stopped at dam overnight. Lyons came up from town with men.

July 18, Men went up to work on dry rool.

July 19, Fred Pitch up from town load supplies. Lyons here with me to take up supplies.

July 20, Rain fall lasting 4 hrs. ½ in. fell.

July 21, Rain fall lasting 6 hrs. ¾ in. fell.

July 22, Sunday.

July 23, 5 million run over dam.

July 24, Everything well. Light showers.

July 25, The pond rose 10 in. water from upper dams.

July 26, Rain fall lasting 2 hrs. ¼ in. fell. Shut booms. Put splash [planks] on big rool.

July 27, Fred Pitch up with load supplies.

July 28, Crew of 13 men sent up to work on rool. Fred Pitch left dam with load blankets. By order from company.

July 29, Cook left. Sent team down for one.

July 30, Took off splash and sluced 5 millions [board feet] of logs. And then put on splash.

July 31, Team belonging to Eggona went down. Fred Pitch up load supplies.

AUGUST 1883

Aug. 1, McLeod up with orders for flood. Rain fall lasting 2 hrs. ¼ in. fell. River rose 8 in. by flood from upper dams.

Aug. 2, Driving crew moved down to Surveyor's. Luk Lyons, K.D. McLeod went to look at rear in Jump River.

Aug. 3, Fred Pitch with load supplies. Team came up to work on rear.

Aug. 4, Light showers lasting 3 hrs. Crew went up River Jump to rool in the Jump [dry roll].

Aug. 5, Commenced flooding. Wasted flood.

Aug. 6, Shut down for to raise head. Pitch up with load supplies for driving.

Aug. 7, Dam shut down. Water rising slow.

Aug. 8, Team and load of supplies went on the dry rool.

Aug. 10, K.D McLeod up with orders to raise the river below dam. Divided the crew on Jump. Some working on the mouth and others at the rear.

Aug. 11, K.D. McLeod went down river. Light showers. Dark and cloudy.

Aug. 12, P.M. raise the river 10 in. below the dam by orders K.D. McLeod.

Aug. 14, Water raised 6 in. in the pond by flood from upper dams. 2 men from dry rool for supplies. Crew came up from town going on Birch Creek for Heller.

Aug. 15, Old Jack left the dam and damn sorry he was.

Aug. 16, Opend booms up at piers and sluced 6 millions.

Aug. 17, Rain fall lasting 4 hrs. ¼ in fell.

Aug. 18, Fred Pitch with load supplies. 3 men from Birch [Creek] for tools.

Aug. 19, Dan McDonald team up with load of supplies for dry rool. Rain fall lasting 3 hrs. ¼ in. fell.

Aug. 20, Luk Lyons down with 17 men to work on jam on Brunies Falls during flood.

Aug. 21, Took off splash and sluiced 5 millions of [board feet] logs. Rain fall lasting 1 hr. Flooded full to ___(?).

Aug. 22, Dam shut down. Helers crew returned from Birch Creek

Aug. 24, Luk Lyons and crew working on the falls below dam. Ran wanigans from falls to mouth of Fisher.

Aug. 25, Little Falls team went to town for orders. Ran

wanigans of[f] Brunies Falls.
Aug. 26, Shut up the dam by orders K.D. McLeod.
Aug. 27, Rain fall lasting 2 hrs. ¼ in. fell.
Aug. 29, K.D. McLeod with order flood.
Aug. 30, Driving crew mooving kit over Brunies Falls. John Whidden up to boss the dry rool from Brunie Falls down [to Chippewa Falls]. Luk Lyons went up Jump with men and teams to put in rear.
Aug. 31, Opened six gates to flood. Whidden over Brunies Falls. Rain fall lasting 3 hrs. ¼ in fell. Full head on dam set for flood.

SEPTEMBER 1883

Sept. 1, Opened booms and sluiced 6 millions. Order stop flood.
Sept. 4, Water in pond rose 6 in. by floods from upper dams. Little Falls team whent down for orders John Whidden run down (?).
Sept. 5, Luk Lyons received orders 5 in. below the dam.
Sept. 6, Rain fall 4 hrs. ½ in fell.
Sept. 7, Shut up dam by orders.
Sept. 8, Stiles team up from (?) for a load of blankets and tents to be used on Fair Grounds.
Sept. 9, Light snows. Heavy dews.
Sept 10, Fall rain lasting 6 hrs. ½ in. fell.
Sept. 11, Two men down for supplies for crew on Jump.
Sept. 13, Light showers.
Sept. 14, Fred Pitch load supplies.
Sept. 15, Rain fall lasting 4 hrs. ½ in. fell.
Sept. 16, Flood from Jump River raised pond 4 in.
Sept. 17, Full head of water. 5 mill[ion feet of logs] in pond.
Sept. 18, Team left for orders.
Sept. 19, Team returned. Orders to flood.
Sept 20, Rain fall lasting 8 hrs. ½ in. fell.
Sept. 21, Rain fall lasting 2 hrs. ¼ in. fell. Two millions logs from Jump.
Sept. 22, Rain fall lasting 12 hrs. ½ in. fell.
Sept. 23, Rain fall lasting 6 hrs. ¼ in fell.
Sept. 24, Opened booms. Sluiced 8 millions. Flood.
Sept. 25, Team from Little Falls [with] hay. All well. Bright and clear.
Sept. 26, Shut down gates to rais water in Jump. Wm. Anderson stoped here. Looking land for D. Shaw.
Sept. 27, All right.
Sept. 28, Wanigan down from rear for hay.
Sept. 29, Luk Lyons took his oxen to work on rear.
Sept. 30, Sluiced 5 millions over dam.

OCTOBER 1883

Oct. 1, Luk Lyons went up to look at rear in Jump.
Oct. 2, Shut down everything close to raise head by orders K.D. McLeod.
Oct. 3, Fred Pitch load supplies.
Oct. 4, Water raising fast.
Oct. 5, All well. Clear and bright. Rain fall lasting 8 hrs. ½ in. fell.
Oct. 6, Tow men down from Jump for supplies.
Oct. 7, K.D. McLeod with orders to flood and also to stop Jump drive or dry rool.
Oct. 8, Crew stoped dry rooling. Brought wanigans and kit down to dam. Rain fall lasting 8 hrs. ¾ in. fell. Flood.
Oct. 9, Part of crew to work getting rock for winter purposes. The remainder on the road. Total 21 men.
Oct. 10, Luk Lyons team to work on road. Sent all hired teams down.
Oct. 12, Repairing blacksmith shop.
Oct. 13, Rain fall lasting 9 hrs. 1½ in. fell.
Oct. 14, Rains and sunshines. Well such is life. Seven million [feet of] logs cam into booms.
Oct. 15, Shut dam down to rais head to sluice. 9 millions logs came to jam piers. Flood from Jump rais pond 5 in.
Oct. 15, Sluiced into dam.
Oct. 17, Team from John Whidden. Wanigan up with load. Damage hash or chuck provisions if you wish. Rain fall last 6 hrs. ½ in. fell. Word up from K.D. McLeod.
Oct. 18, Draw the pond six inches in six hrs. Sluiced ten millions.
Oct. 19, Sluiced fifteen millions. Pat Hines kit arrived from [Elm] Bottoms, charge of Boardman.
Oct. 20, Crew busy moving kit over Little Falls. Also all of the driving tools from Little Falls.
Oct. 21, Boardman left Little Falls with all Hines driving kit. Also all driving implements from Little Falls. Rain fall lasting 4 hrs. ½ in. fell.
Oct. 22, Sluiced 6 millions down to the pond.
Oct. 23, Sluiced 15 millions over dam. John Whidden stoped over night at Little Falls.
Oct. 24, Two teams up with feed & lumber.
Oct. 24, John Whidden left Little Falls with 4 yoke oxen.
Oct. 25, Preparing to clear trees out of pond.
Oct. 27, A man up from town with orders to flood the twenty eight 28th.
Oct. 28, Flood 12.30 P.M. Bigst flood of season.
Oct. 29, Rain fall lasting 5 hrs. 4 in. fell. Run wanigans over dam. Tied up below.
Oct. 30, Shut down gates to pick rock.
Oct. 31, Wanigans left for Chippewa Falls,

NOVEMBER 1883

Nov. 1, Sent a man over to Pat Murray's camp after blasting tools. But found none.
Nov. 3, Opened thirteen gates to give water to run wanigans.
Nov. 4, Sunday. Luk Lyons returned from town to flood.
Nov. 5, Rain fall lasting 9 hrs. 1 in. fell. Shut down gates. Rais head for flood.
Nov. 7, Sent two men to put all boats in pond by orders K.D. McLeod. Full head on dam. Sent word down.
Nov. 8, Man returned. Orders to flood.
Nov. 9, Thos Murphy left here with six boats. Sluiced six millions logs.
Nov. 10, 2 millions logs came in to booms.
Nov. 11, Flood, 11.30 P.M. 7 feet 6 in.
Nov. 12, Let the water run natural hight of river.
Nov. 13, Frose up for the season.

APRIL 1884

April 15, Sluicing 16 hrs. Flood 11.30 P.M.
April 16, Shut down 28 gates.
April 17, Water raising fast in river.
April 18, Full head on dam. Opened booms and sluiced four hours.
April 19, Sluiced 16 hrs.
April 20, Sluiced 16 hrs.
April 21, Sluiced 16 hrs.
April 22, Little Falls team to town for load.
April 23, Sluiced 16 hrs.
April 24, Sluiced 16 hrs.
April 25, Sluiced 16 hrs.
April 26, Sluiced 16 hrs.
April 27, K.D. McLeod up with orders for flood. Rain fall lasting 5 hrs. ½ in. fell.
April 28, Light showers, Sluiced 16 hrs. and flood 12 P.M.
April 29, Shut down 28 gates. Crewing (?) working on jam below dam. Rain fall lasting 9 hrs. ½ in. fell. Shut dam

all down for fresh head.
April 30, Shut down for fresh head.

MAY 1884

May 1, Water rising fast.
May 2, Full head on dam.
May 3, Opened booms. Sluiced 14 hrs.
May 4, Sluiced 16 hrs. Rain fall lasting 14 hrs. ¾ in. fell.
May 5, Sluiced 16 hrs. Team went to town for load.
May 6, Sluiced 16 hrs.
May 7, Sluiced 16 hrs. Team returned load supplies.
May 8, Louis Misheler paid his men here for driving on Jump and went down. Pond rose 10 in. Flood from Jump. Sluiced 16 hrs.
May 9, Sluiced 16 hrs.
May 10, Sluiced 16 hrs.
May 11, Sluiced 16 hrs.
May 12, Sluiced 16 hrs. and shut booms. By orders K.D. McLeod.
May 13, Crew to work on road.
May 14, Water falling fast.
May 15, Shut down all gates.
May 16, Put on splash.
May 17, Sluiced 16 millions into the pond. K.D. McLeod up with orders flood the 20th.
May 18, Team whent to town.
May 20, Flood 12 P.M. Sluice 25 millions, Sent five to work on jam Jump Islands.
May 21, Rain fall lasting 6 hrs. ¾ in. fell.
May 22, Rain fall lasting 9 hrs. 1 in. fell.
May 23, Full head on dam.
May 24, Sluiced 16 hrs.
May 25, Sluiced 16 hrs.
May 26, Sluiced 16 hrs. Shut booms.
May 27, Crew to work on road.
May 28, Crew to work on road.
May 29, Water falling.
May 30, Rain fall lasting 4 hrs. ¼ in. fell.
May 31, Light showers.

JUNE 1884

June 1, Crew to work on road.
June 2, Crew up and opend booms. Sluiced 10 hours.
June 3, Sluiced 16 hrs.
June 4, Sluiced 16 hrs.
June 5, Crew went up to Flambeau to pull jam.
June 6, Came back to work on road.
June 7, Rain fall lasting 6 hrs. ½ in. fell.
June 8, Rain fall lasting 5 hrs. ¼ in.
June 9, K.D. McLeod up with orders to sluice 50 millions and draw pond down 4 feet.
June 10, Sluiced 24 hrs.
June 11, Sluiced 24 hrs.
June 12, Crew back to work on road.
June 13, Rain fall lasting 7 hrs. 4 in. fell.
June 14, K.D. McLeod with orders. Sluiced 14 hrs.
June 15, Sluiced 16 hrs.
June 16, Sluiced 16 hrs.
June 17, Sluiced 16 hrs. Flood, Rain fall lasting 4 hrs. ¼ in. fell.
June 18, Crew working on road. Rain fall lasting 5 hrs. ¼ in. fell. Jump River crew working on road.
June 19, Rain fall lasting 5 hrs. ½ in. fell.
June 20, Crew of 20 men up from town to work on jam at Flambeau Farm.
June 21, Rain fall lasting 4 hrs. ¼ in. fell.
June 22, Crew up from labor on road. Took off splash and sluiced 4 hrs.
June 23, K.D. McLeod with orders for flood. Sluiced 16 hrs. Rain fall lasting 6 hrs. ¼ in. fell.
June 24, Team gone down for load. Sluiced 16 hrs.
June 25, Big jam in Flambeau and also from jam piers up to 4 miles in Flambeau River.
June 26, Jam pulled rear of jam at the mouth of Flambeau.
June 27, Luk Lyons came down and commenced sluicing.
June 28, Sluiced 16 hrs.
June 29, Sluiced 20 hrs.
June 30, Everything going up river. Head wind.

JULY 1884

July 1, Stoped sluicing and prepared for flood.
July 2, Flooded. Waste flood. Part of crew went up on drive at Flambeau to put in rear on Flambeau River.
July 3, Received orders to cut down the wages.
July 4, Shut down for fresh head.
July 6, Rear all down to mouth of Flambeau. Pulled jam at Jump Islands.
July 8, Moved wanigans down and pulled jam over rapids at Flambeau Farm.
July 12, Crew puting in rear. Watter very low. Dam all shut down for fresh head.
July 14, Flood 12 P.M.
July 15, Ran boats down to wanigan and prepared to suspend dry rool.
July 16, Water very low. Paid all the crew at Flambeau Farm.
July 17, Went down to town.
July 18, Ingram up from town. Little Falls crew stoped working on road. Team up from town. K.D. McLeod up with orders to break roolways at Brunies Falls.
July 24, Water very low.
July 25, Letting water go natural run of river.
July 28, Roolways all brokin. Men all paid up and went to town. Business suspended. No men at Little Falls. By orders from general headquarters.

AUGUST 1884

Aug. 1, Nothing doing.
Aug. 25, Business commencing.
Aug. 26, Edw. Johnson came up from town.
Aug. 27, Team up from town with load.
Aug. 28, Commenced to sluice.
Aug. 29, Sluiced 12 hours.
Aug. 30, Sluiced 12 hours.
Aug. 31, Sluiced 18 hours. Flood. Let water run natural of river.

SEPTEMBER 1884

Sept. 2, Men to work on road cutting out road up to Flambeau Farm.
Sept. 10, Noting of any imporance.
Sept. 11, Men to work on road. Luk Lyons left for Oshkosh.
Sept. 15, Monday night heavy rain fall lasting 8 hours. Rain was heavy they seen before.
Sept. 16, Heavey rain fall. All night water rising like a tide. All the splash taken off and water still rising.
Sept. 17, All gates up. Every pice splash taken off and everything don to let water escape and still rising in the pond. About five P.M. o'clock the dam went out. The far side bursting first about half an hour before this side. The logs piled up on the dam, blocked up the gates and bust the booms taking some of the piers with them.
Sept. 18, Water falling. Men gathering up stuff and puting in warehouse. Bridges have all gone out and other piers badly dammaged. Booms are all broken. Part of

warehouse taken away. Blacksmith shop and old stable all gone.
Sept. 19, Men getting stuff picked up and stored away.
Sept. 20, Preparing to work on road and clear out logs, build bridges, cut rock.
Sept. 21, Men working on road. Logs running down river.
Sept. 22, Tryed to shut booms at jam piers.
Sept. 23, Water to hight and current to strong to close booms,
Sept. 24, Logs still coming.
Sept. 25, Men and team working on road.
Sept. 26, Crew working at Bobs Creek Bridge. Met Catharth's crew 2 miles below Bobs Creek.
Sept. 27, Bridge finished and men working up to Little Falls. Men working on Brunuys Hill road all finished.
Sept. 29, Men preparing to build dam.

OCTOBER 1884

Oct. 1, Commenced to clean logs off the dam. Men all paid up till first October.
Oct. ?, Men working at dam clearyin away logs.
Oct. 4, 5 men came up from town looking for work. The crew is getting larger all the time.
Oct. 8, Got all the logs cleared away from the south end of dam and are clearing away the bottom.
Oct. 9, Putting up blacksmith shop. Some cutting timber for coffer dam.
Oct. 10, Commenced sawing plank with Armstrong's mill.
Oct. 11, Got foundation of dam laid down and some planking commenced.
Oct. 12, Arangements made to ship lumber from Big Bend to build dam and hose for men to stay in while building dam.
Oct. 20, Arangments made to have Wm. England locat dam. Water to high to see or sound bottom of river yet.
Oct. 26, Commenced to build house for men.
Oct. 27, Foundation all laid of hous.

NOVEMBER 1884

Nov. 3, England with crew came up today. Men to work gathering rock.
Nov. 14, K.D. McLeod came here and arranged about stuff that would have been needed.
Nov. 5, Men cutting timber for dam and some wheeling rock into the works at south end.
Nov. 30, Report sent down for the month. Men working at dam.

DECEMBER 1884

Dec. 1, Got the foundation of works laid nearly across except small gap in center. Building large butment or pier at the north.
Dec. 10, Wm. England is down geting iron rails, cars and other stuff for running gravel. River ice is getting good, men are commenced hauling rock today.
Dec. 13, Got 265 loads of rock in dam.
Dec. 20, Another week by and 956 loads of rock in.
Dec. 21, Sunday. Luk Lyons has gone to town on business, and I have caught up to date which I will endeavor to keep to from this day forward.
Dec. 22, 3 teams left here for Big Bend for load supplies. Pat O'Neal up from town with load of iron for track. Went back today. Bigest days haul of rock that was don yet 138 loads, six teams.
Dec. 23, Preparing carpenter shop for blacksmith forge. Luke Lyons came up from town. 3 teams came here with load from Big Bend.
Dec. 24, Some men gone down to spend Christmas. Teams went up to Big Bend for more supplies.
Dec. 25, Christmas Day. Not much done to day. Blacksmith moved in the new shop. Two teams came down with load supplies. Weather fine and frosty.
Dec. 26, Team went down to town for load. Robt Murray to work with crew clearing away and repairing rool on the N. side.
Dec. 27, Pat O'Neal up with load of iron for track. Also Saturday today and 662 loads of rock were hauled in with 6 teams this week making in all up to now 1612 loads.
Dec. 28, Sunday. Men coming up from town. Also load of iron.
Dec. 29, Bridge was formed across to haul stone into the east side of dam. 15 loads stone hauled in today.
Dec. 30, Commence clearing logs of piers so as to build and repair them up. Team went down for supplies.

JANUARY 1885

Jan. 1, Load came up from town. Weather cold and frosty. One man hurt by bank of ground falling on him in gravel pit.
Jan. 2, Two teams went to town for load supplies, flour and feed.
Jan. 3, McLeod came up from town and meeting was hel to find out what was wanted here.
Jan. 4, One team came up from town with load supplies.
Jan. 5, Team up from town with load supplies. And went down again. Commenced hauling gravel out of bar on west side or river.
Jan. 6, Ed Johnson and men finishing piers on this side and commenced building piers on the rool. Mike Haley with men reparing piers in pond. Chris Gormerly building pile driver and nearly completed. Cap Henry stayed here overnight on his way up river.
Jan. 7. Chris Taylor came up from town with load supplies and went down for another load.
Jan. 8, Team up with load iron.
Jan. 9, Jo Hogan up with load feed and went down to town for another load.
Jan. 10, Chris Taylor went down to town for load this morning. Load of feed. Mr. Weyerhauser and Mr. Ervine [Irvine] are here today.
Jan. 11, Mr. Weyerhauser and Mr. Ervine returned to the Falls. Seem pleased with all the works and made calculations on send[ing] 40 thousand feet of plank for dam.
Jan 12, Monday. William England left here to oversee some work on Bruney River to be absent about ten days. Some team that were here have got loose last night and chawed each other up.
Jan. 13, Tuesday. 3 teams came up from town with loads supplies. Chris Gormerly and crew to work cutting piles for jam piers. 2 teams went to the Falls for load supplies.
Jan. 14, J. Hogan who was toting remained up to haul piles.
Jan. 15, Jo Hogan went to town for load of supplies. Chris Taylor and another team came up from town with load supplies. K.D. McLeod came here tonight.
Jan. 16, K.D. McLeod went up river today and in the afternoon went down to town. ___? Grant came back to work today noon. Alex Le Tourneau left here today with his team.
Jan. 17, 485 loads rock hauled up today bring this amt. for the week. Also 231 loads gravel.
Jan. 19, Monday. Men all to work and all well. Chris Taylor up from town with load supplies and went down to town for another load. Pat O'Neal hurt today by stone

from a shot in the quarray.
Jan. 20, Tuesday. Teams up from town with load supplies and went down for load of fresh meat.
Jan. 21, All well. Robert Murray got pier on this west side of dam nearly completed. Rock going in fast.
Jan. 22, Commenced to working at piles on the ice and some teams hauling out square timbers for dam.
Jan. 23, Tote team up from town with load feed.
Jan. 24, Men all to work as usual.
Jan 25, Men all resting today.
Jan 26, Monday and the savage performance has commenced once more. Lots of teams here now.
Jan. 27, K.D. McLeod came here today with Mr. Reed who is looking over the books and seeing if everything is all ok. Teams came up with load supplies. Chris Ellis and Pat Wall and went back for another load.
Jan. 28, Wm. Reed is busy today and his business never has come out satisfactory. Mr. K.D. McLeod has been look over some logs up river.
Jan. 29, Mr. Reed and Mr. McLeod have start homeward going over to Pinegards mill and then to town.
Jan. 30, All well. Reduced the crew to 155 men having sent away 15 men, 19 teams to work now.
Jan. 31, Mr. Lyons and Wm. England are going down to town tonight.

FEBRUARY 1885

Feb. 1, Monday. The amt. of rock hauled last week was 725 loads making up to date 4,523 loads in all.
Feb. 2, Men to work driving piles.
Feb. 3, Christ Taylor and Pat Wall went down for load iron and feed.
Feb. 4, Men busy planking dam.
Feb. 5, Chris Ellis and Pat Wall up from town with load iron and feed and went down again taking two batteaus with them.
Feb. 7, Received some hay from Flambeau Farm.
Feb. 8, Christ Ellis went to town for load supplies.
Feb. 9, Men to work putting rock into the piers in pond.
Feb. 10, Mr. Lyons made his wife a present of a horse today.
Feb. 11, Had a visit from Mr. Stafford, ticket agent for the Eau Claire or American Hospital Aid Association.
Feb. 12, Christ Ellis and Patrick Wall up from town with load supplies and went down for another load feed.
Feb. 13, All well and work going on rapidly.
Feb. 14, Teams up from town with load iron, and went back for load supplies.
Feb. 16, Cap Henry, Mr. Ingram and Mr. Chinn (?) were here tonight. Mr. Weyerhausar and Mr. Douglas were here on the 15th. They are all going up river.
Feb. 17, Two teams up from town with load supplies and went down for load iron.
Feb. 18, Crew to work at west end of dam putting plank on. James Hatley here with load of lumber from Alx. Pinegard mill.
Feb. 19, Weather turning warmer and roads are good.
Feb. 20, Ed Johnson and crew putting on string at west end of dam.
Feb. 21, Piers are gettin well fill with rock. There is now 6468 loads in.
Feb. 22, Sunday. All well and K.D. McLeod was here tonight on his way down river.
Feb. 23, Pat Wall up from load feed and down for load flour.
Feb. 24, Lots of plank being hauled here to plank piers.

MARCH 1885

March 1, Sunday and all well.
March 3, K.D. McLeod up from town.
March 4, Robt Murray and crew comenced building cobwork [cribwork?] to close cap [gap?] in dam at east side.
March 5, Water rose in the pond 12 inches. Gormerly commenced capping piers.
March 7, Murray stoped working at gap in dam account of water rising in pond and stopped men from working on pile piers.
March 8, Men to work building booms in pond.
March 10, Rosu (?) caps on pile piers today. K.D. McLeod was here tonight.
March 11, K.D. McLeod went down to town. Luke Lyons went over to Pinegards mill today. Men in quarry stopped working on the 9th [of] Mar. and were paid up and sent to town.
March 12, Total amt of rock put in piers and dam for this season was eight thousand two hundred and 37 loads.
March 13, Weather warm and snow going fast. Ice very thin on Pond.
March 16, Booms are going rappidly and will soon be completed. Teams went for load supplies.
March 17, Teams went to town for load supplies.
March 18, Finished driving piles today.
March 19, Sent down 6 teams and two men gradually reducing the crew now. Mr. Culvor and Mr. Arnold were here today.
March 20, Men to work taking logs off and clearing away from the gates. 4 teams up from town today with load supplies.
March 21, Two teams went down for load supplies.
March 23, Finished clearing logs off the dam and 17 men sent down. Two loads supplies received and went for more loads.
March 24, Finished pile piers all to covering. Weather fine and warm.
March 25, Men to work finishing building gap in dam. Water rising in pond.
March 26, Water rose 3 feet in pond.
March 27, Weather fine and warm.
March 28, Finished the pile piers complete.
March 29, Sunday. Men to work on dam.
March 31, Work going on rappidly. Wm. England left here today for Bruney River.

APRIL 1885

April 1, Weather cold and look like a storm. Thunderstorm in afternoon.
April 2, Weather turning colder.
April 3, Crew to work putting in toe piling.
April 4, Piling all in and everything stoped. Water rising about 2 feet running through the gates.
April 5, Sunday. Weather fine and warm.
April 6, Monday. Men to work butting splash plank.
April 7, Election of town officers and this is named the town of Cleveland.
April 8, K.D. McLeod up making observations of the works.
April 9, Went down to town.
April 10, Logs and drift stuff coming in the gates. Robert Murray and crew to work clearing this out.
April 11, Commenced graveling end of pier on the east side of dam.
April 12, Sunday and men all going down on acout. of low wages. Robert Murray left here today. Team came up from town with a waggon, the first for the season.

April 13, Monday and all is well. Water falling.
(Here the keeper of the Journal becomes confused in his dating and starts over from Aprril 10.)
April 10, Dam completed and men all paid up for winter work.
April 11, Ice thrawing fast and booms are being moved to their place.
April 12, Sunday. Some logs coming in the gates.
April 14, Received orders to pay the men up to the tenth of this month.
April 17, Water rising in pond.
April 18, Ice is all gone this morning.
April 19, Water rose partly by a flood from Deer Tail. Logs are running fast and there are fifteen millions in booms.
April 20, Sluice some old logs and stuff that was in the pond through the rool on this west side of dam.
April 21, Water rising.
April 22, Water rising and all gates up.
April 23, Got the Brave on the bridge.
(Here the journal jumps to June 15th and the entries appear in a new handwriting.)
June 15, All hands sluicing. Water raising.
June 16, Some men went to break a jam on Jump, but could not with low water.
June 19, Some men went up river looking for jams, one on Jump.
June 20, Men went up on Jump to break jam but had no water.
June 21, Rain fall last night most of the night. Logs are running thick. Comenced sluicing after dinner sluiced through small sluiceway 12 hrs.
June 23, Got through sluicing this morning.
June 22, Toat[tote] team gone down for load lumber.
June 24, Pat Hines crew were here today and left their wanigan kit.
June 25, Mesers. Weyerhauser and McLeod were here yesterday and went down today.
June 26, Toat team went down for load feed & oats.
June 27, Kit taken from here today.
June 28, Toat team came up. Brought orders from Mr. McLeod to flood Monday night at 2 o'clock.
June 29, Sluiced tonight, and then flooded.
June 30, Seventen drivers went down river today.
(At this point in the Journal a hiatus exists and there is no entry until the following year when the "remarks" open on April 15, 1886, in a new handwriting. The man who filled in during this break may have kept a separate Journal.)

APRIL 1886

April 15, H. Gannon start to town with order for men start sluicing.
April 16, H. Gannon back with order from K. McLeod to take care of Jump from the mouth of Little Jump. 9 men from town.
April 17, Man gone to Bobs Creek to see if they want men. Water to high to sluice.
April 18, K. McLeod came up yesterday. Went back today. 10 men gone to Bobs Creek to drive.
April 19, Man went to rear of jam. Logs coming in very slow.
April 22, 3 men down from Jump. Rear of jam at Jacco's Island. Logs coming in fast.
April 24, Team went to town. 10 men from Bobs Creek. Sluicing very slow. Head winds.
April 25, Let ten men go today. Logs running good.
April 16, Rainfall last night ¾ in. All hands braking jam at Jacco's Island. Team up from town with load of supplies.
April 27, Mr. Meagher from Bobs Creek here today. Mike Hayes went down with team to fetch up kit. Sluiced last night.
April 28, All hands at Jacco's Island breaking jam.
April 30, McLeod came up today. Jam broke today at Jacco's Island.
(On this page at least three different handwritings appear from May 1 to 16.)

MAY 1886

May 1, McLeod went to town today. Everything in good shape on Jump River.
May 2, Good haul on sluicing. Light shower today.
May 3, Good sluicing all day.
May 4, Light shower last night 1/8 in. fell.
May 5, Head wind every day.
May 6, Man from John England drive rear at Brinks place.
May 7, Later. Accounts from Jump Rear at Shoulder Creek. Man down after tobacco from England's drive.
May 8, Bad wind for sloosing. South west wind.
May 9, Put on all the splash plank, and shut down all the gates. Heavy rain all night fell 1 good inch.
May 11, 3 men from John Englands drive for grub. The Porkey Hog came up today to drive Alder Creek.
May 12, Dull and cloudy. Not many logs running through & slow.
May 13, Tote team from Chippewa Falls. Fine day, fair wind, but no logs running. Men from England's drive for more grub.
May 14, Booms all closed.
May 15, Men picking out old jam at jam piers. Very cold wind north. Orders from below for a flood.
May 16, K. McLeod here today. Tote team from the Falls.
May 17, Commenced sluicing. Heavy shower with wind from N.W. John England here from Jump drive.
May 18, Three men from England's drive for more grub. Orders to flood.
May 19, Very warm. Flood tonight.
May 20, Shut down the gates for another head of water. Men cutting out of slough at jam piers.
May 21, Water raising fast in pond.
May 22, Wind N.W. Very warm. Water raising in pond.
May 23, Commenced sluicing. Thunder showers.
May 24, Head wind this afternoon. Tote team left for town.
May 25, Jam at Brunett Falls. Men down breaking it.
May 26, Got jam off at Brunett. Commencing sluicing again. Tote team from town.
May 27, K. McLeod up with order to flood.
May 28, Let off a flood. Shut down gates again.
May 29, Very fine wind. N.W.
May 30, Water raising in pond.

JUNE 1886

June 1, Commenced sluicing again. Raining a little all day. Fell 1¾ inches.
June 2, Fair wind. Logs running good.
June 3, Jam on Brunett Falls. Men down breaking it. Jam all clear again.
June 4, K. McLeod up today. Put on splash boards. Water raising pretty fast. Thunder showers from north.
June 5, 6 men gone to Jims Falls to break jams. Water raising in pond.
June 6, Wind west. Sluiced what logs there was in pond. Flood tonight.
June 7, Put on all the splash boards. 6 men went to town.

June 8, Cloudy. Water raising slow. Tote team left for town.
June 9, Wind N.W.
June 10, Tote team from town. Very warm today. Water raising slow.
June 11, Thunder showers last night lasted about 4 hours. Very warm. K. McLeod up today.
June 12, Water raising.
June 13, ½ inch rain last night. K. McLeod up today. Water came up fast last night.
June 14, Heavy rain this evening.
June 15, Orders to flood tonight. Good water today. Flood.
June 16, Tot team gon to town. (This entry is made in different handwriting.)
June 17, Men working pol taxes. McLeod up today. (This entry same handwriting as above. Handwriting in next reverts to June 15th style.)
June 20, Sunday. Wind from the S.W. and nice growing showers. Men working up in the pond.
June 21, Sluced 8 hrs. Jamed on the Falls and men worked on it and put her through. [Probably Brunet Falls].
June 25, 19 gates up. Water running natural run of river. Men to work cutting snags in pond.

JULY 1886

July 1, Water falling fast in pond and river.
July 3, Shut gates to raise a head in pond. Small rain fall.
July 4, Part of the crew went down to town.
July 5, Water rising slow in pond. Weather very hot and no rain. Wind south west.
July 6, Men to work polling logs out of sloughs in the pond.
July 10, Run wanigans down to Jims Falls.
July 12, Crew preparing and loading wanigan to work up in the pond beginning at Flambeau Farm.
July 13, Team went to Flambeau Farm to dry role. All hands to work in pond.
July 19, Heavy shours last night.
July 20, Shoury last night. Began to cut hay.
July 26, 6 men cutting trees in pond. 5 cutting hay.
July 29, Toteam went to town.
July 30, Tot team up with load suplies.

AUGUST 1886

Aug. 2, Eight men cutting trees in pond. 3 men graveling.
Aug. 4, J. Whidden & 6 men staid over night.
Aug. 7. Whidden & crew staid for one day & half.
Aug. 10, Toteam went to town after load supplies.
Aug. 13, Dispatch for to shut down.
Aug. 15, Flood 7 foot one.
Aug. 20, Heavy rain fell 1.42 inch last night.
Aug. 21, 2 inches of rain last night fell. ½ inch rain today.
Aug. 24, Toteam home from town with load of supplies. Dispatch to catch a head of watter.
Aug. 27, Heavy rain up river. None heare of eny act.
Aug. 29, Sent man down town with dispatch.
Aug. 31, North west wind every day bad wind for our work.

SEPTEMBER 1886

Sept. 1, McLeod and Whidden here over night.
Sept. 3, We have the dead wood on the old gam [dam?] on the island all OK.
Sept. 7, Whidden & 4 men came after a wanigan. Light shours but donte amount to eney thing.
Sept. 8, Tote team gone down to Falls. Cloudy, looks like rain.
Sept. 9, Light showers all day. Water not raising any.
Sept. 10, Fine with northwest wind. Tote team from Chippewa Falls. Commenced digging potatoes.
Sept. 11, Light rain last night. Cloudy today.
Sept. 12, Water raising slow.
Sept. 13, Men & team hauling out piles at jam piers.
Sept. 16, Heavy rain last night fell 4¾ inches. Heaviest rain for 21 years. Sent man to the Falls with dispatch. Water raising.
Sept. 17, Wind west. Very cold. Water still holding up.
Sept. 18, Commenced raining this morning.
Sept 19, Water falling.
Sept. 20, Orders to commence sluicing. Shut down gates & took of splash.
Sept 21, Good water for sluicing. Not very good wind.
Sept. 23, Head wind. No sluicing today. All the gates down. Orders from town for 2 feet water. Raised 8½ gates for 2 feet.
Sept. 25, Not very good sluicing. Head wind. Showery all day. McLeod up with orders for flood.
Sept. 26, Logs running good.
Sept. 28, Sent down a d-m good flood.
Sept 29, Tote team left for town. Men gravelling the dam.
Sept. 30, Cold N.W. wind. Water raising slow.

OCTOBER 1886

Oct. 1, Tote team from town. The Rev. John Dodge paid us a visit.
Oct. 2, Pete Bolen came down with wannigan & driving kit.
Oct. 3, Whidden arrived with driving kit & took most of it down with him.
Oct. 4, Logs running slow. Tote team left for town.
Oct. 5, K. McLeod up with orders. Closed boom and put on splash planks.
Oct. 6, To catch head. Flood tonight 11 P.M.
Oct. 8, Tote team from town with load supplies.
Oct. 10, Old horses from dry roll came up.
Oct. 11, Sorrel team left here.
Oct. 12, Team hauling hay from meddow. Heavy rain fell yesterday and last night. Water raised some in river.
Oct. 13, Water raising fast.
Oct. 14, Cold N.W. wind.
Oct. 20, Team up from town with load of supplies.
Oct. 24, Heavy rain last night. Ran two wanigans down below the dam.
Oct. 26, Tote team from town with load & orders for flood. Put on splash & shut down gates.
Oct. 24, Water raising pretty fast.

NOVEMBER 1886

Nov. 4, Water falling in pond.
Nov. 6, First snow of season, Turning cold.
Nov. 9, Five men here for dinner. Let them have 14 pairs blankets by order from K. McLeod.
Nov. 13, Water commenced falling. Very frosty nights.
Nov. 15, Tote team left for Chippewa Falls.
Nov. 16, Snow storm. Very heavy fall.
Nov. 17, Tote team from town with load. Ordered to take off splash & let the water run off slow.
Nov. 24, (This entry scratched, two lines.)
Nov. 25, Tote team left for the Falls.
Nov. 27, Team from town with supplies.

DECEMBER 1886

Dec. 1, Orders for all the horses at the dam to go to town.
Dec. 22, Gormlys team from town with load.
Dec. 27, Tote team went to town.
Dec. 29, Tote team arrived from town with supplies. (On opposite page recording water levels at the dam this notation at bottom: "Record closed for the season.")

APRIL 1887

April 8, Water commenced raising. Had to raise some gates. Wind S.W.
April 9, Very warm. Ice going out of the pond fast.
April 10, Water raising fast. Ice near all gone.
April 11, Some rain fell last night. Showery all night. Ice all out of pond. Sent a man to town for men. Orders from town not to sluice for a few days.
April 12, Very warm. Water raising yet.
April 14, Water raising. Took splash off one roll on other end of dam. Three men from town to drive.
April 18, Put on splash on small roll. Orders from town to commence sluicing.
April 19, Commenced slucing this morning. Logs running slow.
April 21, Wind N.W. Logs running pretty good. 13 men arrived to work on logs.
April 22, Heavy snow storm, the worst of the season.
April 23, K. McLeod here today. Logs jamed at Birch Creek. Fell 12 inches of snow yesterday. Pretty good for the 23 of April.
April 29, Five of McLeods farmers went down. They could not go on logs. So they had to take a walk.
April 30, Sluicing slow by orders from below. Some rain last night. Water running in river.

MAY 1, 1887

May 1, Water raising fast. Wind S.W. Very warm. Tote team left for town. Logs coming in fast from up river.
May 2, Heavy rain last night. Water still raising. Wind S.W. blow a gale.
May 3, Tote teams from town with supplies. Water on a stand [?]. Sent team down with man that was killed with Ligtning on John England's drive.
May 4, Water fell a little last night. Commenced sluicing again.
May 5, Logs running good. Sluicing day & night.
May 8, No 1 here today on a visit to to see the dam. Orders for a flood tomorrow at 10 o'clock A.M.
May 11, John England & men here from Jump drive. Brought kit down. Drive almost out. Rear at Devils Nest.
May 12, Very warm. Signs of rain. Water falling fast.
May 13, Jam all through. Logs running through. Orders for a flood.
May 14, Logs running pretty good.
May 15, Flood tonight 12 P.M.
May 16, Tote team went to town.
May 18, Team arrived from town with supplies.
May 19, K. McLeod up with orders. Shut down all the gates. Water raising slow.
May 20, Put on splash planks to catch a head. Tote team gone to town with potatoes.
May 22, Wind S.W. Water falling in river.
May 23, Flood tonight. 1.30 P.M. [Probably an error for A.M.]
May 24, Cloudy & cool wind N.W. Water falling fast.
May 27, K. McLeod up with orders to stop the work. Water too low.
May 28, Tote team went down this morning. Tote team arrived from town.
May 29, A good rain last night. Rained heavy today with heavy hail. Showery all day.
May 30, Showery. Watter raising slow. No logs comming in.
May 31, Logs coming in pond. Cloud. No rain. Watter at a stand [still?]

JUNE 1887

June 2, Water raising. Sluiced some logs. Man went to town with dispatch.
June 3, No. 1 here today. Flood wanted.
June 5, Flood tomorrow morning 1 o'clock A.M.
June 8, Thunder shower last night.
June 13, Heavy showers last night. 1 inch rain fall.
June 14, Rain fell 1 inch last night.
June 15, McLeod up with orders shut down for more water.
June 17, Water raising slow.
June 18, Thunder shower this afternoon.
June 19, Commenced sluicing today. Very warm.
June 20, Sluicing very good. Water raising. Ten men commenced work. Showery all day.

June 23, Orders to flood tonight.
June 24, Flood last night.
June 25, Shut down & put on splash planks to catch another head.
June 27, Water raising in pond. Tote team gone to town.
June 28, Tote team arrived from town. Orders for a flood. Shut down. Water raising slow.

JULY 1887

July 3, Showers all day.
July 5, Orders for a flood tomorrow morning 1 A.M.
July 11, Tote team gone to town. Commenced making hay.
July 15, Shut down all the gates.
July 16, Water raising slow. Very hot weather.
July 19, Sluicing through jam piers. Logs running slow. No water from up river yet. Looks like rain.
July 21, K. McLeod here looking for water from Shugar Bush dam. None to be seen. 1½ inch raise in river.
July 25, Commenced to sluice. Water falling in pond. No water coming in. Very hot.
July 26, Randolph Smith went to town. Crew making hay.
July 27, Heavy shower today.
July 28, Cap Henry's team went down. Looks like rain.

AUGUST 1887

Aug. 2, Showery all day.
Aug. 3, Water falling in pond. Everything shut.
Aug. 4, Tote team gone to town.
Aug. 5, Orders to waste what water is in pond and fix gate.
Aug. 9, Showery this afternoon.
Aug. 10, Repairing gate in dam. Heavy day's rain from the south.
Aug. 11, Water very low in river.
Aug. 12, Man gone to town with dispatch.
Aug. 13, Raining all day from south. Water not raising.
Aug. 14, Very dark day. Looks like rain.
Aug. 16, Tote team gone to town.
Aug. 19, Tote team arrived from town.
Aug. 21, Team gone to town.
Aug. 22, Heavy showers last night. Raining all day. No sign of a raise yet.
Aug 23, Crew started to Flambeau to dry roll. Clear wind. North.
Aug. 24, K. McLeod here. Shut down to try and catch a head.
Aug. 30, K. McLeod here today looking for water.
Aug. 31, Commenced raining this morning. Showery all day.

SEPTEMBER 1887

Sept. 1, Heavy rain last night & this morning. Water raising in river. Took off splash & commenced sluicing. Man gone to town with dispatch.

Sept, 2, Head wind. Logs running slow.
Sept. 3, Cloudy. Looks like rain. Logs running good.
Sept. 4, Orders for a flood tonight. Cloudy with light showers.
Sept. 5, Sent down a good flood.
Sept. 6, Men working on dam at Brunett Falls.
Sept. 7, John Whidden up for two wanniggan boats. Water raising fast in pond.
Sept. 8, Head wind not very good sluicing.
Sept. 9, Tote team from town.
Sept. 11, Head wind. Jam almost through.
Sept. 12, All the logs through the jam piers. Man gone to town with dispatch.
Sept. 13, Flood tonight 1 A.M.
Sept. 19, Dandy is away heart broken.
Sept. 20, Heavy thunder shower.
Sept. 21, Ordered to shut down & catch another head. Shut down gates & put on splash.
Sept. 27, McLeod here with orders for a flood.

OCTOBER 1887

Oct. 2, Showery with heavy wind from S.W.
Oct. 4, Tote team gone to town.
Oct. 6, Orders to shut down & catch a head.
Oct. 7, Heavy rain last night.
Oct. 8, Cloudy with light showers.
Oct. 11, Man gone to town with dispatch. N.W. wind. Very cold.
Oct. 12, Heavy wind from S.W. Full head on dam.
Oct. 13, Old Doc arrived from town with dispatch. Orders for a flood.
Oct. 16, Flood tonight 11 A.M. Light rain falling all day.
Oct. 17, Water very low in pond.
Oct. 18, Smith went to town with tote team. Tom Murphy here picking up boats. Took 8 from the dam.
Oct. 20, Tote teams from town with orders to shut down & catch another head.
Oct. 23, First snow storm of the season. Fell 3 inches. Man gone to town with dispatch.
Oct. 24, Colder than Hell.
Oct. 26, Orders to waste the head. No flood wanted.
Oct. 31, Tote team gone to town with Jump driving kit.

NOVEMBER 1887

Nov. 3, Tote team gone down with another load. Smith went to town with tooth ache. [Probably Archie Smith, later a foreman on the drive.]
Nov. 7, Tote team to town with load.
Nov. 12, K. McLeod here for a small flood.
Nov. 14, Working at the old Parker gate.
Nov. 17, 2 men of Joe Wiler heare today for supplies land hunters.
Nov. 27, Man from town for all the old horses. Dam cold weather. About 6 inches of snow on the ground. "Record closed for the season."

APRIL 1888

April 9, Water commenced raising in river.
April 10, Snow going away fast.
April 11, Men cutting ice from pond. Wind N.W. Clear.
April 12, Cool wind. South.
April 13, Warm wind from S.W.
April 14, Very warm. Ice melting out of pond.
April 15, Wind N.W.
April 17, Heavy rain last night with some hail & snow. A channel through the ice from the piers to the dam.
April 18, Ice all out of the pond. Water too high for sluicing. Cool N.W. wind.
April 20, Put on splash plank on small rool and shut down some gates. Preparing to sluice. Man gone to town with dispatch for men. Not many logs in pond but some coming in. Wind S.W.
April 22, Big jam on Chippewa at Surveyor's Point. Man gone to town with dispatch. Crew working on jam.
April 23, Man from town with dispatch.
April 24, Jam all broke down. Pond full of logs up to Joco Island. No orders to sluice yet.
April 26, Thunder showers this morning. Crew working on Jump. Wind S.W.
April 27, Raining all last night. Water raising fast. All the gates up. Took splash off small roll this morning, and opened another roll at far end of the dam. Team could not pull them so had to trip them.
April 28, Man gone to town with dispatch. Opened another roll this morning. Water still raising. Raining all last night. Fell 1½ inch. 12 A.M. opened another roll. 4 P.M. opened roll on this end. 9 P.M. water still raising. Opened another roll. Snowing. Wind north.
April 29, 2 A.M. commenced taking off another roll. Boom broke at jam piers. Some logs running. 6 A.M. water about a stand. 11 A.M. water raising. Commenced to take off more plank. 7 P.M. Some logs came down and got into the gates. 12 P.M. everything open that will let water escape.
April 30, 1 A.M. Water commencing to fall a little. 2 P.M. Water going down.

MAY 1, 1888

May 1, Water falling. Heavy frost last night.
May 2, Heavy rain this afternoon. Water still going down. Men down river picking up splash planks.
May 3, K. McLeod here. Raining all day.
May 4, 10 men went down to clean out the slough at Fishers [creek]. Rain.
May 7, Commenced putting on splash again. Rain again.
May 9, Raining all day. Water raising.
May 10, Heavy rain this afternoon.
May 11, Orders to get ready for sluicing. 12 men from town to work on logs. Commenced putting on splash.
May 12, Finished putting in plank and shut down some gates. Water very high in river.
May 13, Commenced sluicing.
May 14, Crews picking on jam at piers.
May 15, K. McLeod here today. Not many logs running. Jam working very bad. All the crew picking on jam.
May 16, About seven hours good sluicing.
May 17, Men breaking jam all day. Very hard to get logs. Two men gone down to Carfields to boat up supplies. All the bridges gone and cant get a team up. 4 men from Jump drive for supplies.
May 18, Tote team with load of supplies. Light rain all day. Logs running slow.
May 19, Team hauling up plank for the dam. Hank Moffet and No. 2 here. Jam working bad. Four-horse team arrived with a load of plank.
May 20, Two teams arrived with plank. No haul on the jam yet.
May 21, 58 men from town to work on jam. Heavy rains last night.
May 22, Men working on jam got it broke down from Jump. Water raising.
May 23, Heavy rain this morning. Water too high to sluice.
May 24, Water too high to sluice.
May 25, Jump drive down. Forty men from dam to take Jump kit and work on jam at Surveyor's Point.
May 26, Commenced to break jam at Surveyor's Point. Two

teams arrived with supplies.
May 27, Water falling. Commenced sluicing 2 P.M.
May 28, Very good sluicing. Hank Mofett here.
May 29, Good sluicing today.
May 31, Logs running good. Water fallin in river.

JUNE 1888

June 1, Let off a small flood in order to get all the splash planks on. All the rolls closed & gates shut down for another head.
June 2, All the crew gone up to Joco Island to break jam.
June 3, 8 men came down to sluice.
June 4, Head wind. Bad sluicing. Crew still working on jam at Joco Island.
June 5, Head wind today. Logs running slow. S.W. Chinn here looking for logs.
June 6, Water falling in river. Good sluicing.
June 7, Jam on Brunett Falls. Crew went down to break it. 12 P.M. jam all right. Logs running good.
June 8, Let of[f] flood this morning. 6 P.M. shut down to catch another head. Got a big haul on the jam at Flambeau too.
June 9, Commenced sluicing 12 o'clock A.M.
June 11, Toteam arrived from town with load of supplies. Logs running good.
June 13, Heavy thunder shower last night. Water still raising. Logs running good.
June 14, Heavy squall from N.W. last night. Logs running good.
June 20, Heavy shower from S.W. this morning. Water holding up rear of jam at Joco's Island.
June 21, Heavy showers yesterday. Water raising some.
June 22, Rear of jam down to Birch Creek.
June 23, No. 1 & No. 2 here today orders to stop sluicing. Got enough logs below for a while. Wages cut down. 24 men left for town.
June 27, Sluiced what logs was below the jam piers.
June 28, Very warm. Water falling in river.
June 30, Water falling fast. Very hot weather.

JULY 1888

July 1, Tote team from town with supplies.
July 4, Reparing roll where logs had torn it out. 4th of July. Very lonesome here today. Dance tonight at the 9 Mile House but too far for us to go.
July 5, Randolph went to Flambeau for a wannigan.
July 6, Brought wannigans & kit down to Jump Islands to commence work in the pond.
July 9, Commenced work in pond.
July 10, Orders for a flood tonight 12 P.M.
July 13, Smith went to town.
July 15, Tote team from town with supplies.
July 16, Man gone to town with dispatch about water.
July 17, Orders for a flood tomorrow night.
July 19, Flood last night.
July 20, Commenced to make hay at Lebay's.
July 21, Heavy squall from west.
July 25, Water raising very slow in pond.
July 28, Heavy showers this morning.
July 29, Pond full. Orders for a flood tomorrow night.
July 30, Flood tonight 12 o'clock P.M.

AUGUST 1888

Aug. 1, Heavy thunder storm last night.
Aug. 3, Rain last night. Water raising.
Aug. 4, A good rain in the river for this time of year.
Aug. 6, Heavy rain last night. 2 feet raise in the river.
Aug. 7, Cloudy all day. Water still holding up.
Aug. 8, Showary. Wind N.W.
Aug. 9, Water raising in river.
Aug. 10, Cool and cloudy.
Aug. 11, Some rain last night. Cloudy. A man down from the wannigan for vegetables. Sent up all he could carry.
Aug. 13, Cloudy. Wind S.W.
Aug. 14, Commenced sluicing. Water falling. Opened small roll to sluice and shut down some gates.
Aug. 16, Water falling but logs running good.
Aug. 18, Orders to flood tomorrow night.
Aug. 20, Commenced cutting oats. Very warm.
Aug. 25, Very hot weather.
Aug. 27, Finished making hay. Tote team to town for supplies.
Aug. 28, LeFave moved the wannigan to Surrey end Point to work. No water to work on[?] Joco Island.

SEPTEMBER 1888

Sept. 7, Finished dry roll at Flambeau. Teams hauling kit over the jam. All the crew going down.
Sept. 8, Light shower last night. Wind N.W.
Sept. 9, Smith & Gormly arrived here with dry roll kit on their way to the Falls.
Sept. 11, Team hauling kitt over Jim's Falls. Team went on down town.
Sept. 14, Rain all afternoon, not heavy, moderate.
Sept. 15, Some rain last night. Cloudy today. Watter very low in the river. No moove in the watter.
Sept. 20, Crew came down from Jump Islands. All gone down.
Sept. 21, Orders for a small flood.
Sept. 28, Orders for another little flood to be on the 30th, at 6 oclock P.M.

OCTOBER 1888

Oct. 1, Cold N.W. wind. Men digging potatoes. Smith started to town. Tote team gone down for supplies.
Oct. 3, Doc arrived from town with supplies.
Oct. 5, Shut down for another flood.
Oct. 18, Rain from south. Flood tonight.
Oct. 20, 4 men arrived to work on dam. Slight fall of snow.
Oct. 22, Cloudy and d-m cold.
Oct. 23, Commenced work on the apron of the dam.
Oct. 25, McLeod here with orders for another five-cent flood.
Oct. 26, Showery this afternoon.
Oct. 28, Flood tonight 6 P.M.
Oct. 30, Tote team gone to town. Shut down all the gates for another flood.
Oct. 31, Man gone to town with dispatch.

NOVEMBER 1888

Nov. 5, All the men went down to vote for General Harrison.
Nov. 6, Boler & the horse is lost or sold to free traders.
Nov. 7, Republicans feels good now & Democrats is all down in the mouth. J. Jardine is home from election, laughing.
Nov. 8, Orders to flood tonight.
Nov. 15, First snow storm of the season. Fell one inch.
Nov. 16, Colder than Hell and Damnation.
Nov. 17, Pond frozen over.
Nov. 20, Splended weather for working on the apron. Smith gone down the road to hire another team.

MARCH 1889

(A new man has taken over the Journal.)

March 29, Ice nearly all out of the pond. Archay arrived

with eight men to commence work on pond.
April 1, Cold with slight fall of snow.
April 3, K. McLeod here looking for logs & wattr. Shut down gates to try and catch a head.
April 6, McLeod & Ervine here for logs & flood. Opened booms & commenced sluicing.
April 7, Heavy wind. No logs running.
April 8, Flood tonight—a one-horse flood.
April 9, Shut down tonight for another head. Man gone to town with dispatch.
April 10, Man arrived from town.
April 11, Doc started to town.
April 12, Water raising in river.
April 13, Doc arrived with supplies.
April 14, Man with dispatch for logs. Bad sluicing. Head wind.
April 15, Head wind. Logs running slow.
April 16, Head wind. Some logs sluiced last night, but running slow today. Light shower this afternoon.
April 17, Logs run good last night but south wind today. Some logs running.
April 18, Same as yesterday.
April 19, Wind N.W. Logs running like damnation.
April 20, Sent down a good flood this morning.
April 21, Shut down & put on splash.
April 23, Man gone to town with dispatch about water.
April 24, Took off splash & commenced sluicing. Rain & snow. Wind North West. K. McLeod here with 20 men for Archay.
April 26, 27 men from town to work on jam.
April 27, Men rolling in rear. Logs running good.

MAY 1, 1889

May 1, North wind for this last six days.
May 2, Tote team started for town. Very frosty nights. Logs running good.
May 3, Smith started for town this morning. Very fine weather. No rain yet.
May 4, Warm with wind south. Smith arrived from town. Water commencing to fall.
May 5, Orders to cut down the wages & commence dry roll. 24 men left for town.
May 6, K. McLeod for a flood Wednesday.
May 7, Wind south, blowing like hell. 4 P.M. Heavy wind from southwest.
May 8, Flood today. Repairing the leak in the dam. Heavy snowers all day.
May 9, Stopped the leak and shut down the gates.
May 10, 12 A.M. Pond full. Big raise in river.
May 11, Good sluicing. All the jam through piers.
May 12, Few logs running. Squint-eyed Billie got his drive out of Elder creek.
May 13, Logs running right through & cheating the poor drivers. Quite a few logs running.
May 14, Flood this morning 3 o'clock A.M. Toteam went to town.
May 15, Shut down for another head. Cloudy.
May 16, Showery with north wind. Very cold. Crew braking jam at Devils Nest.
May 17, Cludy & cold. Few logs running. Water raising.
May 18, Commenced sluicing. Showery all day. McLeod & Archay down here. Flood Tuesday morning.
May 21, Flood this morning. Tote team arrived. Shut down for head. Water running fast.
May 25, Logs all sluiced. Pond clear.
May 28, Logs running through the piers & cheating the drivers.
May 29, Kenneth McLeod here with orders for a flood the 3rd.
May 30, Jam on Brunett Falls. Crew down breaking it.
May 31, Good flood this morning.

JUNE 1889

[A new hand takes over the Journal in April-May-June, etc. and the writing is delicate but firm, with few misspellings.]
June 1, Dam cold weather. Looks like snow. Shut down for another head.
June 3, Took off splash to commence sluicing.
June 4, Banlay Parnell up here for two wannigans. Tote team started for town to take a load to Flambeau.
June 5, McLeod here for a flood Friday morning.
June 6, Commenced raining this afternoon.
June 7, Flood this morning 3 o'clock A.M. First summer day.
June 8, Shut down for another head.
June 9, Doc from town with supplies.
June 10, Took off splash & commenced sluicing.
June 11, Water falling fast & lots of logs comming in.
June 12, Water still falling.
June 16, A raise in the river today & lots of logs coming in. K. McLeod here today.
June 17, Raised four gates to give a small raise in the river. Bad sluicing today.
June 18, Very warm weather. Water holding up. 15 men started for Devil's Nest.
June 20, K. McLeod here with orders for a flood. Showery all day. Slight raise in river.
June 21, Sent down a flood this morning. 10 P.M. 20th raised three gates for to raise river 1 foot. 12 P.M. 20th raised seven gates to give 1 ft. 6 in more and at 3 A.M. 21st, let the rest go.
June 22, Put on splash & shut down for another head.
June 25, Orders for a flood Friday.
June 26, Took off splash & tried to sluice but head wind and logs wont run good.

JULY 1889

July 2, Very warm. Heavy thunder showers. Flambeau drive is out.
July 4, Man gone to town with dispatch. Very fine day.
July 5, Flood this morning 3 A.M. Smith went to town.
July 7, Smith arrived from town. Water very low in river.
July 8, Light showers. Very warm.
July 11, Man gone to town with dispatch. Quite a raise of water in Jump but cant sluice any for head wind.
July 12, Took off splash & commenced sluicing.
July 13, Orders to send crew from Devils Nest to drive on Jump.
July 15, Tote team started for town.
July 21, Heavy showers this morning from N.W.
July 25, Showers last night. Water falling in pond. Man gone to town with dispatch about water.
July 26, Tote team gone to town.
July 28, Men up for a wannigan for Anderson.
July 29, The Flambeau drive hung up on Brunett Falls with his wannigan.

AUGUST 1889

Aug. 2, Tote team gone to town.
Aug. 8, Tote team gone to town. Heavy showers this morning. K. McLeod here today.
Aug. 9, Water raising a little.
Aug. 10, Showery all day.
Aug. 17, K. McLeod here for a flood. Shut down.

Aug. 20, Flood last night 12 P.M.
Aug. 21, Repairing leak in dam. Slight raise in river. Shut down some gates this evening.
Aug. 22, Water raising fast in pond.
Aug. 24, Pond full. Took off splash and commenced sluicing.
Aug. 26, Sent down a damn good flood & shut down for another head.
Aug. 27, Archie & LeFavie arrived here with dry roll kit.
Aug. 28, They started down with the boats.

SEPTEMBER 1889

Sept. 4, Sent down a good flood. Tote team arrived from town with supplies. Showery all day.
Sept. 11, Toteam went to town. Repairing booms.
Sept. 17, Tote team started for the falls. Heavy N.W. wind & d-m cold.
Sept. 19, Cold as Damnation. Heavy frost last night.
Sept. 24, Showering all day.
Sept. 25, Tote team arrived with Tilbett to cook.
Sept. 26, First snow storm of the season.

OCTOBER 1889

Oct. 2, Tote team went to town. Smith went down this afternoon.
Oct. 3, Very heavy frost.
Oct. 4, Smith & tote team arrived from town.
Oct. 7, Team hauling hay from Beaver Meddow. Very fine weather. Water very low in river.
Oct. 26, S.W. Chinn here looking over the dam. Splended weather, but no signs of rain.

NOVEMBER 1889

Nov. 1, About 2 inches of snow, the first of the season.

APRIL 1890

April 6, Heavy rains. Water commencing to raise.
April 7, Rain with thunder. Wind S.W.
April 8, Water raising like hell.
April 10, Water still on the raise and ice solid in pond.
April 12, Opened boom last night and let a few logs go. Jam on Jump. Man gone to town for men.
April 15, Orders to commence sluicing.
April 17, 16 men from town to work on logs.
April 22, Logs running good. Light rain. Wind south.
April 23, Orders to stop sluicing for three days.
April 26, Crew working on Joco Island poling out logs.
April 27, Commenced sluicing again.
April 28, Logs running good. K. McLeod here with orders to stop sluicing. Water falling in river. About 15 men working on Jump at the Devils Hole.
April 30, Tote team gone to town.

MAY 1890

May 2, John Fleming arrived with running drive from Jump. Water very low.
May 4, K. McLeod here with orders for a flood tomorrow 10 o'clock A.M. Snow storm this afternoon.
May 6, Shut down for another head.
May 11, Shut down for another head. Water very low in river.
May 14, Jump drive hung up. All the men came down.
May 15, Wind N.W. with showers. Good sluicing.
May 16, Heavy N.W. wind. Flood tomorrow morning.
May 18, Light rain last night. Heavy fall of snow last night up river.
May 21, Rain all last night.
May 22, Heavy rain from east.
May 24, Sent down a good flood 1 oclock A.M. Commenced to shut down this evening. Showery.
May 25, Showery all day.
May 26, Water raising like hell.
May 28, Orders for another flood Saturday. 14 men arrived to drive at Devil's Nest but no water there so they had to go down again.

JUNE 1890

June 1, Very warm. Water running fast.
June 2, Showery all day.
June 6, Orders to stop sluicing. Rear of jam at Deer Tail [Creek].
June 10, Showery all day. Man gone to Flambeau to watch the West Fork wannigan.
June 20, Very warm. Logs running slow.
June 21, K. McLeod here with orders for a flood.
June 25, Warm as hell.
June 28, Warm with light rain.

JULY 1890

July 2, Dam full. Commenced sluicing this morning.
July 19, Archey arrived with men to dry roll at the Devil's Nest.
July 22, 4 teams from town to work on dry roll.
July 24, Man gone to town with dispatch. Showery. Wind north.
July 25, Sent down a dam good flood this morning.
July 28, Tote team gone to town this morning for supplies.

AUGUST 1890

Aug. 2, Heavy thunder showers.
Aug. 3, Showery all day.
Aug. 8, Heavy wind. Showering all day.
Aug. 9, River beginning to raise.
Aug. 10, 5 feet raise in the river.
Aug. 11, Toteam gone to town. Jardine started for N.B. Canada.
Aug. 12, Heavy showers last night. Showery all day today. Watter falling in river.
Aug. 13, Toteam from town with oats and with orders to sluice 10,000,000 [feet of] logs for C.L. & B. Co. [Chippewa Lumber & Boom].
Aug. 14, S.W. wind all day.
Aug. 15, Cant sluice any logs today.
Aug. 16, Good wind today. Logs running good.

Here the Journal breaks off with more than half the remaining pages left blank except for these entries on one of the last pages dated Oct. (no year) under the initials CRILD Co which would be the Chippewa River Improvement Log Driving Company.

Oct. 10, 11 men over night and team	3.80
Oct. 11, 36 men dinner and 4 teams	11.80
Oct. 12, 1 man team over night	1.50
Oct. 30, 1 man team over night	1.50
Oct. 31 1 man team over night	1.50
Nov. 2, 1 man team over night	1.50

Brave on the Bridge

In the daily journal kept at the Little Falls dam, an entry for April 23, 1885, makes this terse announcement: "Got the brave on the bridge today." To anyone unfamiliar with the folk lore of Chippewa County, the word "brave" here might be taken as a misspelling for "brace" because, for one thing, the word is written with a lower case "b", and bridges do need braces.

But the person who made the entry in the journal was writing about a sculpture of an Indian made from a big pine log, more than life size, which was mounted on a bridge of the Little Falls dam that day and anchored with ropes (later iron rods) to make sure the winds did not tear him loose.

The Brave faced west because that was the direction the logs on the Chippewa drive were coming from before being sluiced through the Little Falls dam. In the early days of logging, the Chippewa River followed a big curve here as it came down from the north, and it turned east for about two kilometers before it turned south to continue on to Brunet Falls. It was on this lower bend in the river where the big dam was built, the first one in 1878, and the second in 1885. Here the towering Indian with the piercing eyes stood, unfazed by winter snow and summer sun. He was lost in a big flood which tore part of the new dam out but he was rescued, less one arm, which James Jardine remade and grafted onto him.

The creator of the Brave, Luke Lyons, is mentioned several times in the daily journal. There could be a second man in the background who was involved with the sculpture, Jean Juvette, who may have found the original piece of cork pine, that is, pine free of knots, which he sold or gave to Lyons with the suggestion that he make something of it. Obviously, Lyons had done something with a drawshave or jacknife before this time, for it is doubtful whether anyone could have achieved such a fine piece of sculpture unless he had previous experience. He probably worked on it in his spare time from the late 1870s into the 1880s, yet, he was a busy man. From the references to him in the daily journal, it seems that he was one of the straw bosses employed by Chippewa Lumber & Boom, for he was constantly at work around the dam, or on log drives, or as an assistant to Kenneth B. McLeod.

Lyons seems to have left the company in March, 1885, and was not present in April when the Brave was mounted on the bridge.

After the last log drive in 1911, the fate of the Brave remained in doubt. Who was going to watch

Photo of Indian Brave at Little Falls dam taken about 1910. Here Brave has gilded point on spear. Chain mechanism in foreground controls gates on dam. Crib piers in back holds boom which funnels logs into sluice gate.

Close-up view of Indian Brave at Little Falls dam probably taken not long after dam was completed in 1885. Two steel tracks at lower right were used for four-wheeled cart which carried building materials and rocks out to point of construction from south shore of dam.

over him? Youngsters like Zac Jardine, whose father was foreman of the dam crew in the early 1900s, crawled out on the dam to have his picture taken with the Brave. There was something about him that made Zac and his playmates proud to be with him. He had been part of the legend of the great *Chippeway* drive and his presence had touched the lives of nearly everyone in the Little Falls community, even after the drivers, in their stag pants and calked boots, had moved to lumber camps of the Pacific Northwest, or departed for the Big Rock Candy Mountain.

The Little Falls dam was taken over by the Minnesota—Wisconsin Light & Power Company, later known as Northern States Power, a company which built the hydroelectric dam a short distance

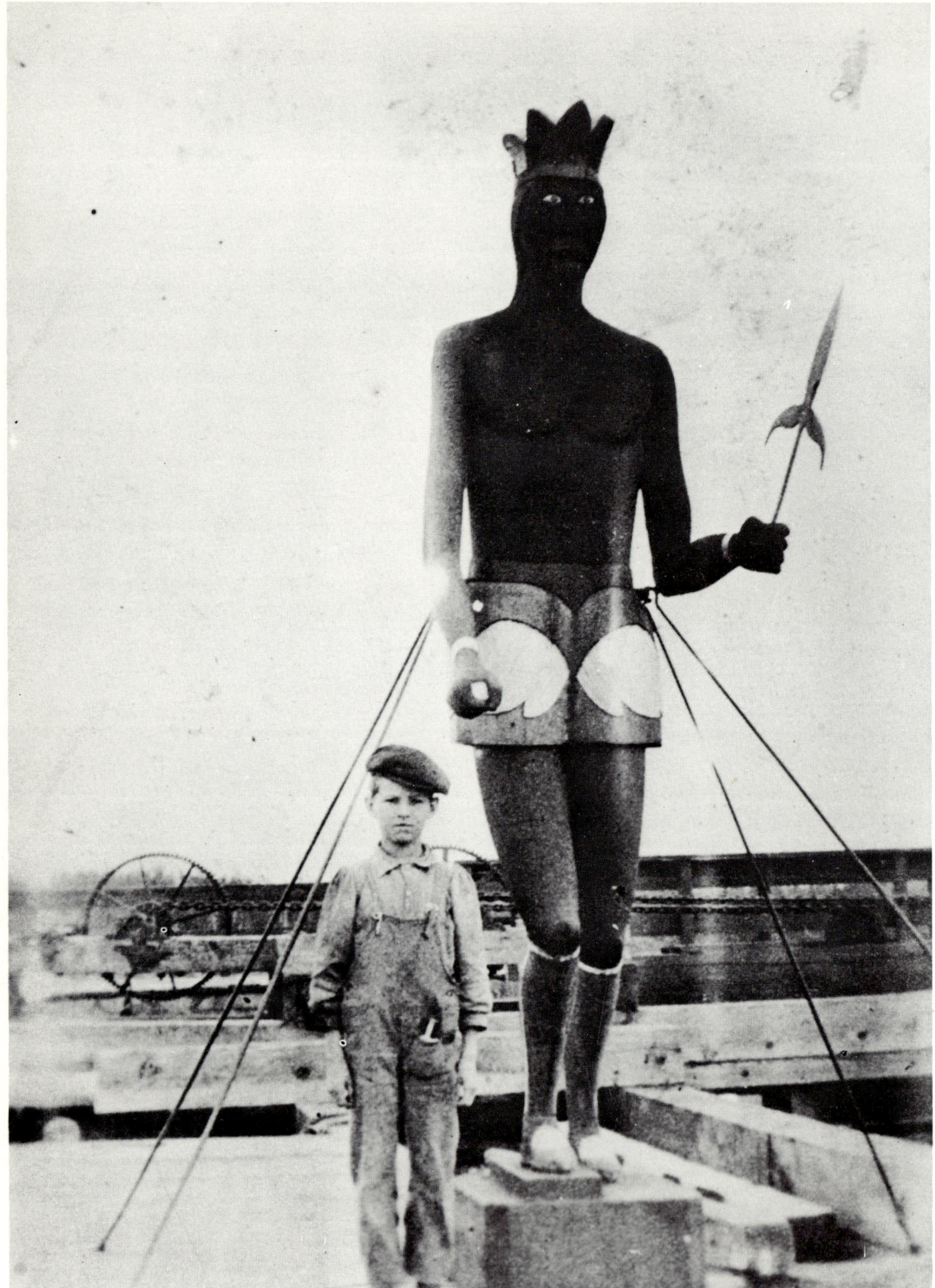

Zac J. Jardine, sling shot in his pocket, had this picture taken with the Brave on the Bridge, circa 1907. At left is part of mechanism probably used for manipulating Tainter gates.

below the one-time log driving dam. With the purchase of the old dam, the company inherited the Brave on the bridge. Fearing for his safety, the company removed him to Eau Claire and kept him in storage. On two occasions he was exhibited at the company's booth at the Chippewa County Fair, and for several months he was on exhibit at the museum of the Chippewa Valley Historical Society in Carson Park.

Citizens of the Little Falls area, fearing that they were going to lose the Brave requested that he be returned to his original environment. The power company agreed, and for a time he was kept in the town hall, but today he stands in a glass case under a partial roof, facing north across the village of Holcombe where he now greets tourists. But whatever became of Luke Lyons? No one seems to know. His memory will not be forgotten, though, not as long as the Brave is guarded in the future as zealously as he has been in the past.

Tragedy at Little Falls

Driving logs on the rivers was always dangerous and many of the drivers lost their lives, but the most tragic accident occurred on July 7, 1905, when eleven men drowned on the Chippewa River. A log jam had developed below the Little Falls Dam (Holcombe) and seventy or more drivers were recruited in Chippewa Falls to break it up. William Irvine, general manager for Chippewa Lumber & Boom Company, told Andrew Gonyea, a driving foreman, to round up as many men as he could find and get them to Little Falls on the morning train. The recruits arrived about 10:30 a.m. the following morning; most were still in street clothes and some had been drinking too much the night before. The train stopped at the Omaha depot on the east side of the river and from there the men walked across the plank road of the dam to get to the west bank where the batteaux were tied.

There are stories still circulating in Chippewa County about the holiday spirit that followed some of these men across the dam, some barely able to walk of their own accord, but all anxious to be the first on the jam, for this was the spirit of competition among log drivers everywhere. They wanted to show the boss how good they were.

When the men crossed the dam, sixteen of them jumped into the first batteau and shoved off from the west bank. The bowman of this batteau tried vainly to hook his peavey into the first log he ran into on the jam, and he might have succeeded had he had the cooperation of the rudder man, Andrew Gonyea, but he was in no condition to respond. As a result the batteau swung 180 degrees in the fast current and the remaining men, who were standing, fell overboard and capsized the boat. Three men, George Kaiser, John Dressle, both of Chippewa Falls, and William Smith of Drywood, had jumped onto the jam at the instant the batteau touched it. Eddie Martin, a seventeen-year old youth, and Emil Toutant, also capsized with the others but had the good luck to be saved. They were picked up a mile downriver.

The rest of the crew, about sixty men, saw the accident from shore, and another batteau with eight men went to the rescue of the three marooned on the jam, but this batteau capsized too, and this time six of the crew managed to cling to the log jam while Eugene Riley and Henry Andrews were swept downstream. They managed to swim ashore and on their return to the dam, they noted that the other members of the crew were standing with their hands in their pockets, frozen by indecision, while the nine men on the jam waited helplessly.

Riley and Andrews jumped into another batteau and made two trips to the log jam to rescue the stranded drivers. The Eau Claire *Leader* said "their bravery had never been surpassed on the Chippewa River," and the Chippewa *Herald* said they were "heroes of the right stuff," an expression revived in the Space Age to describe the Astronauts.

When news of the tragedy reached Chippewa Falls, the big sawmill closed down for the day and Irvine ordered the flag on the mill flown at half mast. Meanwhile, the log jam was ignored and the drivers were ordered to search for the dead. None was found in the first twenty-four hours, and Irvine offered a reward of $50 for the recovery of each body. Farmers along the river joined the search and in a few days all but one of the bodies had been recovered. Andrew Gonyea was not found until August 5th, his body bruised and decomposed. Most accounts agree that he was responsible for the tragedy.

The Chippewa Falls *Daily Independent* on July 8, 1905, (a Saturday) carried a two column headline, ELEVEN LOG DRIVERS DROWN. The newspaper listed the missing men in this order and with these spellings:

Saul Bracket, Eau Claire
Louis Cokey, Flambeau
Max Billiard, Drywood
Paddy Lyden, Stanley
Ole Horne, Chippewa Falls
Joe Pelloquin, Drywood
Andrew Gonyea, Jim Falls
Oscar Berquist, Cadott
Bert Larry, Anson (township)
Henry Ferguson, Chippewa Falls
Adolph Toutant, Cadott

A few days later on July 12th, the *Independent* changed Ferguson's Christian name from Henry to

When Bert Lairy (also as Larry), farmer in Anson township, Chippewa County, drowned, he left wife and five children, four of whom seen in snapshot taken several years after accident. Two boys, left, George and James, were five and seven years old when it happened, and they have grown in experience ahead of their time to help a courageous mother run the farm. Two girls are Anna and Sabina. Right: floral display with photograph of deceased was common at funerals in the 19th and early 20th Century and Bert Lairy's family followed custom.

Ole Horne (upper left) an immigrant from Norway in his youth, became an expert river driver known to his peers as "White Water Ole." He had gone back to Norway for a visit and on his return to Chippewa Falls was married only a short time before tragedy that took his life at Little Falls. Above: Oscar Berquist, an uncle of Joyce Gannon, and lower right, Joe Pelloguin of Drywood, who also drowned.

Byron. In the same issue was an interview with Adolph Bernier, a businessman from Holcombe. Bernier told a reporter that he did not wish to "criticize the dead," but he firmly believed "if it had not been for one or two men in the boat, the accident would not have happened. They [the two men] insisted that it was not overloaded and tried to induce more men into the boat. If, when they reached the jam they had kept their places and not stood up in the boat until a landing could have been effected, things might have been different."

William Hoyer who drove the Chippewa River for many years, was not present at the scene of the accident, but arrived a day or two later. He told the author that several of the drivers were farmers with families who lived along the river. When the body of one of them accidentally drifted ashore near his own farm, and when Whitewater Ole drifted all the way to Chippewa Falls where he lived, a legend grew that the men all drifted ashore next to their own place "as if they were going home," Hoyer said.

Mrs. Henry Ferguson, mother of several children, was unable to cope with the trauma of losing her husband, and on the day of the funeral, July 11, she attempted to take her own life by swallowing a compound of laudanum and ether. A doctor was summoned in time to save her life.

Saul Bracket was working on the river that summer, saving money to continue his education at the University of Wisconsin. He was graduated from Eau Claire High School where he played football.

The story of the drowning of the eleven river drivers was told and retold for many years, and there are people living today around Chippewa Falls, Cadott and Drywood who remember their parents or uncles talk about it, in hushed tones, as though hoping not to offend the fierce river gods who ruled the *Chippeway* drive.

William W. Bartlett of Eau Claire composed a ballad dedicated to Charlie Williams, another driver who drowned and was buried on the banks of the Chippewa River. The ballad "Farewell Angelina" appears in *Ballads & Songs of the Shanty-Boy* (Harvard University Press, 1926).

Eugene Harm holds calked boots and wool socks worn by his uncle, "Paddy" Lyden (top) in Tragedy of Little Falls.

THE RIVER IN THE PINES

By William W. Bartlett

Oh, Mary was a maiden when the birds began to sing.
She was fairer than the blooming rose so early in the spring.
Her thoughts were gay and happy in the morning gay and fine,
For her lover was a river-boy from the River in the Pines.

Now Charlie got married to this Mary in the spring,
When the trees were budding early and the birds began to sing.
"Now, darling, I must leave you in the happiness of love,
And make some V's and X's for you, my darling dove.
And early in the autumn when the fruit is in the wine,
I'll return to you, my darling, from the River in the Pines."

'Twas early in the morning in Wisconsin's dreary clime
When he rode the fatal rapids for that last and fatal time.
They found his body lying on the rocky shores below,
Where the silent water ripples and the whispering cedars blow.

The woodsmen gathered round him on the bright and cloudless morn,
And with sad and tearful eyes they viewed his cold and lifeless form.
"I would send a message to her, but I fear she would repine,"
Spoke a friend of Charlie Williams from the River in the Pines.

When Mary heard these tidings from that river far away,
It was in the early springtime, in the early month of May.
At first she seemed uncertain and no more her eyes did shine,
But her saddened thoughts still wandered to that River in the Pines.

Not long ago I visited there, not many years ago;
It was a Southern city where strange faces come and go.
I spied a gray-haired maiden, both very old and gray,
And my thoughts turned back again once more to that river far away.

She smiled though when she saw me, though she looked old and gray.
"I am waiting for my Charlie boy," these words to me did say.
"And early in the autumn, when the fruit is in the wine,
I'll return to meet my Charlie from the River in the Pines."

Now every raft of lumber that comes down the Chippewa,
There's a lonely grave that's visited by drivers on their way.
They plant wild flowers upon it in the morning fair and fine;
'Tis the grave of Charlie Williams from the River in the Pines.

Keel Boat to Steamboat

Keelboats used on upper Mississippi and Chippewa Rivers were sixty feet or longer. Sketches here show smaller boats used on rivers in the East. Boat at top has four men pushing on poles with one man at rudder. On larger keelboats an equal number of pole pushers were used on both sides of the deck. Sketch below shows keelboat being pulled by *cordelle* (French for "line") method, with men ashore pulling boat over difficult rapids. Line to top of mast gave greater leverage than tied only to bow of boat.

Before railroading came to the United States, most freight between cities and between states was hauled either by ship, wagon, or boat, usually a barge or keel boat, both flat-bottomed and slightly tipped at the bow. The song "Erie Canal" from Carl Sandberg's Songbag come to mind, for it was along the Erie Canal from Albany to Buffalo that this type of carrier is best remembered in American literature.

Neither the Chippewa nor the Sioux Indians had ever heard of a keel boat; they used birchbark canoes to ascend the Chippewa River, and when northern Wisconsin was opened to lumbermen, the white man bought canoes from the Indians and followed them up the river to a falls named after the Chippewa nation. But there was a limit to what a canoe could carry; it was not built to haul freight or mill machinery. The answer lay in the keel boat which could be built by local carpenters from locally sawed lumber, or from lumber sawed with whipsaws.

The usual method on rivers in the East and along the Ohio was to use long poles to push the keel boat upstream by manpower. A running board about two feet wide extended the length of the boat on both sides of the hull. Taking his place at the bow on the running board, the first pole man sank his tamarack pole into the river bottom and started walking backwards while pushing on his

The *Chippewa*, one of most popular steamboats on Chippewa River, was used for several years on Chippewa Falls to Wabasha run. She may have been built in Pittsburg, arriving on Mississippi River in 1854, or even as late as 1858. Ball seen below flag on jackstaff was called "nighthawk." Usually made of metal, it was raised or lowered by manipulating lines only faintly seen on picture. According to Ruth Ferris, former curator of River Room, Missouri Historical Society, nighthawk was used by pilot as sighting device to hold boat on course. It was also used to gauge heights and distances.

pole, often bent over to a forty-five degree angle. Little wonder this was sometimes called "Tamarack power."

Bruno Vinette, the French-Canadian who came to the United States in 1853, went up the Mississippi River in 1855 to find work in the Pinery. He says he rode a keel boat from Read's Landing to Chippewa Falls, and he also agreed to help push the boat. It was back-breaking work, but young men soon got accustomed to it. Vinette said he made several more trips by keel boat up the Chippewa River in 1855.

Talking with William W. Bartlett many years later, Vinette said the keel boats on the Chippewa were sixty feet long, ten feet wide and drew four feet of water. George Merrick, author of *Old Times on the Upper Mississippi*, describes the keel boat before the era of steamboating as the most efficient mode of travel and transporation on the Ohio and Mississippi rivers. There were other types of boats such as arks, broadhorns, scows and flatboats which floated downstream easy enough but which were impossible to push upstream. The keel boat, with upturned ends, fore and aft, was sixty to eighty feet long, and fifteen to eighteen feet wide, with four feet depth of hold. Midship was a cargo box about four feet high. At the ends of the boat was a deck of eight to ten feet, the forcasle deck having a windlass for pulling the boat off bars, or warping through swift water on a curve or rapids.

It was customary for lumber rafting crews to tow a keel boat behind the lumber raft going downriver, and, on arrival at Read's Landing the crew rode the keel boat back either to Dunnville or to Eau Claire. Difficult as poling a keel boat was, it beat walking back through the woods where there were little but Indian trails to follow.

In addition to long poles, the keel boat captains also relied on the *cordelle*, a line nearly a thousand feet long tied to the top of the mast at the center of the boat to a height of nearly thirty feet. The boat was pulled on this line by men walking on shore.

The first keel boat on which Vinette ascended the river carried supplies for H. (for Hiram) S. Allen, a pioneer mill operator at Chippewa Falls. Vinette found a place to live on the south bank of the river in a community called, "Frenchtown," now South Chippewa Falls, where there was a colony of French-Canadians and Indians of French-Canadian descent. He remembered Louis Demarie, a fur trader, a "powerful man, good natured (and) fearless." His wife was the local midwife and their eldest daughter, by all accounts a beauty, married H.S. Allen. In later years Vinette operated a stopping place about twenty-five kilometers above Flambeau Farm on the trail followed by the lumberjacks on the right (west) bank of the Flambeau. It burned in 1875 and was rebuilt. Vinette was also logging in this area for many years.

In the early years, Chippewa rafts were taken as far as Nelson's Landing where they were coupled together to make larger rafts called "Mississippi rafts" for the voyage downstream.

Some time, apparently in the 1860s, the coupling grounds were shifted from Nelson's Landing to the other side of the Mississippi at Read's Landing. From here the returning crews loaded their keel boats with the raft kit in addition to supplies for Knapp, Stout and cast off early in the morning. By nightfall, the keel boat had usually reached Three Mile Prairie lying about ten kilometers (6.5 miles) south of Arkansaw in Pepin County. Here the crew went ashore and put up a tent for the night and cooked supper. It usually took a day and a half to reach Dunnville, but if the stage of water favored the keel boat it was possible to reach Dunnville in one day.

The first keel boat to serve the public as well as Knapp, Stout and Company was *Dutch Lady*, built by Miles Durand Prindle, then of Chippewa Falls, who later gave his name to a community called Durand. *Dutch Lady* operated several years until steamboats largely replaced the need for keel boats.

The first step in the evolution of transporation on the Chippewa came with the introduction of steamboats, although in the beginning, the steamboats were too large for any river as shallow as the Chippewa. In fact, up to 1863, no one had even seriously considered using a steamboat to push a lumber raft on the Mississippi, let alone the Chippewa. Up to now, the great Mississippi rafts, assembled from several Chippewa rafts, were *floated* downriver by oarsmen steering the long oars fore and aft. They could not row the raft; their task was to guide it as much as possible. Finally, on September 12, 1863, a small side wheeler, *Union*, took a raft of 190 cribs and cast off from Read's Landing for Hannibal, Missouri. The raft had a double crew on the bow and when the steamboat broke down below Wabasha, half the crew shipped the stern oars and the raft floated to a point below Keokuk, Iowa, where a packet pushed it to Hannibal.

The following year, *Union* took another raft downriver and this time completed the first successful delivery of a raft on the Mississippi River below Lake Pepin. *Union* was not built for towing, i.e. pushing, big rafts, and about this time, a ship builder, Samuel R. VanSant, developed a stern wheeler with more power and maneuverability.

The *Clyde*, built at Dubuque, Iowa, for Ingram & Kennedy of Eau Claire, was ninety-four ton side wheeler with geared double engines. Her distinguishing feature was iron-plated hull, first on upper Mississippi, built to withstand underwater hazards and floating logs. She arrived at Durand on maiden voyage August 12, 1870, en route to Eau Claire. But company used her mostly for pushing lumber rafts on Mississippi River. In this picture, probably taken at Durand, she may have discharged cargo of breaking plows seen in foreground. Note nighthawk on jackstaff at left.

Within a short time it was found that a big raft could be steered from the stern as well as from the bow, and this was done by lashing a small steamboat across the bow, or front end, called the "bowboat." With one steamboat pushing the raft from the stern, the bowboat, held firmly to the front end, steamed ahead or went astern as ordered by toots on the whistle from the steamboat captain astern. Within a few years there were at least seventy stern wheelers pushing rafts on the Mississippi and St. Croix rivers.

There is no agreement on the first steamboat to ascend the Chippewa River. One source says a charter ran up for John H. Knapp in the spring of 1848 or 1849, and the name of the boat may have been *Dr. Franklin I.* A story in the Eau Claire *Free Press*, under date of July 11, 1867, says that the first steamboat up the Chippewa, the *Otter*, was chartered by H.S. Allen in 1851. Other sources say that a stern wheeler called the *H.S. Allen*, built in 1851, ran up the river that year. The *Otter*, meanwhile saw considerable service on the Minnesota River and as a rafter on the Mississippi, got caught in an ice jam in the winter of 1879-80 and was crushed. She sank into the mud and many years later her captain fished out the boiler and presented it to New Ulm, Minnesota.

The *Free Press* story, mentioned above, says that the *H.S. Allen* was built in 1854 and "ran for two years." Wherever she was built, the *Allen* did not stay long in service on the Chippewa. She was apparently top-heavy and totally unsuited to the shallow river. George Merrick says that he served on the *Allen* about 1862 when she was running between Prescott and St. Paul.

The *Free Press* story referred to here also said that a steamboat called the *Oscota* was purchased by H.S. Allen in 1853 and "ran up" the river five

The *Helen Mar* tied alongside somewhere on Mississippi. Her owners, "Knapp, Stout & Co., Company" here used official name by which company was incorporated. Bottom of sign reads: "Wholesale Lumbermen, Wisconsin & Missouri." Skiff hangs over bow to take captain ashore, or to be used as life boat. In foreground are barrel kegs. Note big night light over skiff. The *Helen Mar* was built in 1872 and died of old age in 1904.

The caption seen in lower left appears on original photograph. The *Annie Girdon* was built for Knapp, Stout & Company and was used for pushing rafts on Mississippi and from Dunnville to Read's Landing on Mississippi. The year 1869 marked first when steamboats were successfully used to push lumber rafts down the Mississippi. Note jackstaff with light for navigation at night. Daniel Davidson was the captain of *Annie Girdon* on this voyage.

or six times in the summer. No other source investigated makes any mention of the *Oscota.*

Fred A. Bill, a steamboat veteran, believes that three boats called *New St. Croix*, *Red Wing* and *Martha Branch*, ascended the Chippewa in 1855, but the *Red Wing* mentioned by Bill was a second boat of the same name and smaller than the first one built in Pennsylvania in 1846. All three of the boats were back on the river in 1856 and business was so good they could scarcely keep up with demand for more cargo space. By 1857 the Waumandee *Herald*, under date of May 9th, had this to say:

> The Chippewa trade has commenced this season with more than usual briskness and our business men here are having their hands full putting up and forwarding goods . . . . Every steamboat brings large additions and our levee is continuously jammed with goods, so that it is difficult to move around among them . . . Keel boating commenced on the 18th of April and was carried on extensively until the river got too high for that class of boat . . . . The steamer *Revellie* is now plying regularly between Read's Landing and the various landings on the Chippewa. She started on her first trip to Dunnville on Tuesday the 5th, loaded with freight for Knapp, Stout & Company, the most extensive lumber manufacturers in the region, and we may say in the northwest.

By 1866, Read's Landing was the busiest port between Dubuque, Iowa, and St. Paul. Noting all the undelivered cargo lying along the levee, one

River steamboats tied alongside at Read's Landing, circa 1870, most of them rafters on Chippewa and Mississippi rivers. Left to right, *Hiram Price*, *L.W. Crane*, *Annie Girdon*, *L.W. Barden*, *Buckeye*, *Clyde*, *St. Croix*, *Wm. Hyde Clark*, and *Silas Wright*. Part of Chippewa raft in foreground serves as jetty.

local wag described the village as "two miles long, two miles high and two rods wide."

Other steamboats which appeared on the Chippewa in 1857 included the *Greenlee* and the *Jennie Whipple*, the latter built in Pennsylvania, but neither boat seems to have spent much time on the Chippewa River.

Between 1840 and 1866, three steamboats on the Mississippi were named *Chippewa*, the first apparently built at Pittsburgh, Pennsylvania, in 1840, for Hercules Dousman, agent for the American Fur Company at Prairie du Chien, and Hiram H. Griffith. The boat worked the Mississippi for three seasons and then vanished, her fate presently unknown.

The second *Chippewa*, according to William J. Petersen, author of *Steamboating on the Upper Mississippi*, arrived at St. Paul from Pittsburgh in 1857, and entered the Chippewa River trade in 1858. In 1859 she was chartered by the American Fur Company to haul supplies to Fort Benton on the Missouri River, a community which, up to this time, had never seen anything but keel boats. But navigation on the Missouri was hazardous, and *Chippewa* got within fifteen miles of Fort Benton on July 17, 1859, discharged her cargo to another boat, and turned back. At this time, says Petersen, she was 3560 miles from the ocean (actually the Gulf of Mexico), a point farther from the gulf by continuous water route than any other boat had ever been.

People in times past were just as eager to set records or to be first as they are today, and *Chippewa's* ascent of the Missouri in 1859 was hailed by one captain as "one of the celebrated feats in steamboat navigation." It sounds rather tame today, but the Missouri in 1859 was a wild

Old bridge at Durand, called "Leonard Bridge," was located about block west, or downriver, from present Highway 10 bridge. Draw operated on wheels to open and allow passing steamboats through. Here *Phil Scheckel* pushes Chippewa raft of three strings of lumber. Long oars at bow are not manned. Crew was probably told by photographer to go ashore.

river, untamed by dams, wing dams or jetties. In 1860 *Chippewa* went back to the Missouri and this time reached Fort Benton. While the boat was tied alongside, the crew was having supper one evening when the cry of "fire" rang out. The boat was carrying gunpowder and the crew cast the boat adrift and watched it explode downstream. The fire was caused by a careless crewman who had gone into the hold to steal liquor.

In 1866, a third *Chippewa* was built, this one at La Crosse, Wisconsin. A boat of only seventy-four tons, she served as a tri-weekly packet plying between La Crosse and Eau Claire, probably reaching Eau Claire on her first voyage in 1866. On June 6, 1867, *Chippewa* the third was being advertised in the Eau Claire *Free Press* as the "fastest on (the) river." After the railroad reached Eau Claire in the summer of 1870, *Chippewa* like most of the other steamboats, could no longer compete for freight and she spent her last season pushing rafts on the Mississippi, and in 1871, while wintering at Rumsey's Landing below Eau Claire, she caught fire and burned to the water line.

In the winter of 1856-57 a boat named *Chippewa Valley* was built by Charles M. Whitney for E.S. Brockway, George Buffington and F.W. Moore of Eau Claire, with Buffington serving as her first master. She was 125 feet long with a twenty-eight foot beam, a rather large boat for the Chippewa River. She was used mostly on the Mississippi and during the Civil War, was allegedly captured by Confederate forces in the siege of Vicksburg.

The *Chippewa Falls*, a stern wheeler built at Pittsburgh, in 1857, probably ascended the Chippewa River in 1859, and was acquired by Knapp, Stout & Company for Andrew Tainter who served as his own master. For a time she was making a round trip a day between Menomonie and Read's Landing and made fifty trips up the river that year. In 1864 she joined a flotilla of steamboats transporting troops up the Missouri River for General Alfred Sully, and that seems to be the last ever heard of her, too.

In 1858, *Red Wing* was back on the Eau Claire-La Crosse run, and a steamer called *Reserve* made her first appearance in Eau Claire, apparently

The *Ida Campbell* on levee, Eau Claire West Side, next to warehouse of Ingram & Kennedy Lumber Company (left). Fourth Street runs at left up to two-story building with tower, probably school house. *Ida Campbell* was built at Durand in 1867 for Ed Campbell who named boat after daughter (later Mrs. Henry Fay). In 1870 Dan Helman and Matt Harris are listed as owners of boat. She wintered one year at mouth of Eau Claire River and was destroyed by fire and later rebuilt, apparently last steamboat to dock at Eau Claire, about 1882. A powerful stern wheeler of thirty-eight tons, she was one of the best built boats on Chippewa River. In lower right, another picture of the *Ida Campbell*, taking on or discharging cargo, probably at Rumsey's Landing below Eau Claire. In foreground over-length logs have been painted with the number "23" to serve as end mark. These painted marks were as legal as marks made with stamp hammer.

Caption appears on original photograph which Union Lumbering Company probably was using for publicity purposes. Picture shows *Buckeye* pushing raft which, like *Annie Girdon* shown earlier, is without bowboat to serve as "point." Men shipping long oars at bow have built wanigan on board for passage downriver, thirteen days to St. Louis. Several oarsmen have built a lean-to to shade themselves against glare of August sun.

'CHIPPEWA FALLS UNION LUMBERING CO

Towed by the Steamer "Buckeye," and run by George Winans from Read's La
August 17th to August 30th, 1869: Containing .... feet of Lumber, 412,000
"FIFTEEN STRINGS, SIXTEEN LONG," covering, by actual measurement, an area of 250
ever run on the Mississippi River.

BILL & CHAMPION,
Owners.

STEAMER "BUCKEYE.'

a "wild boat" (tramp) which never returned.

In late October, 1858, the *Skipper*, a small stern wheeler, Captain E.S. Brockway, came up the Chippewa River. She laid up and spent the winter at the mouth of the Eau Claire River. In March the following spring, she was hit by a heavy wind which tore off her cabin and she sank.

A new boat to the river in 1859 was the *Colonna*, drawing only ten inches of water and built especially for the Chippewa trade. She made forty-two trips up the river in her first season and sank the following spring.

The *Eolian*, a stern wheeler of 106 tons, owned by the Davidson family of La Crosse, probably came up the river in 1859 with eight tons of cargo. She made four trips and on her return from the last one, she turned up the Mississippi River into Lake Pepin, rammed a big chunk of ice, and sank.

The *G.M. Wilson* was running from Read's to Menomonie by this time for Knapp, Stout & Company, and probably made one trip up the river as far as Chippewa Falls, a distance then computed at fifty-six miles (ninety kilometers), in a little more than three hours.

No less than 160,000 bushels of wheat were shipped out of the Chippewa Valley in 1860, mostly by steamboat. The *Chippewa Valley* came up the river in April, 1861, and took on a cargo of 4400 bushels of wheat from the West Side levee in Eau Claire. She left at 8 a.m. and arrived in La Crosse the same evening in time for passengers to connect with a train for Milwaukee.

Knapp, Stout & Company purchased a new boat called *Maquoketa City*, a stern wheeler which they bought in 1861 for $2500. She made at least one trip as far as Chippewa Falls that season.

The *Stella Whipple* was built at Eau Claire in 1861 and later saw brief service as a blockade runner in the siege of Vicksburg. On September 6, 1861, she left Eau Claire with Company C, 8th Wisconsin Volunteers, en route to La Crosse accompanied by an eagle hatched on the Flambeau River and known as "Old Abe."

The *Union* made her debut in 1861, too. On July 19th, the Eau Claire *Free Press* said this about her: "A new steamboat for the Chippewa River has just been built at Waubeek. She is a side wheeler

Interior of excursion boat, probably the *Manitou*. Boat was tied to bow of *Red Wing* (facing page). Lighting in ceiling probably gas; porcelain spitoons and tables probably for poker players.

and bears the name Union."

Union arrived at Read's Landing on her maiden voyage on July 27th, and probably saw more service than any other boat on the Chippewa River. She drew only twelve inches of water, and was able to navigate the river when other boats lay idle.

River pilots were never lacking in stories about their ability to navigate a boat in low water, but the man who drew the least water was probably Louis Malin who was on the *Ida Campbell* one time in 1874 and tied up at the levee in Durand just long enough to tell a local newspaper editor that the water was so low he had passed a number of catfish "of the mammoth variety, aground on the way up." The *Ida Campbell* was advertised in the Eau Claire *Free Press* on June 13, 1867, shortly after she was launched.

For several years in the late 1860s, *Union*, which was owned by Pound, Halbert & Company of Chippewa Falls, was chartered by Captain George Winans at seven dollars a day. He used her for pushing lumber rafts down the Mississippi, and apparently made good money. With his savings he built a hotel in Chippewa Falls costing $40,000. The hotel burned and he had no insurance. He left for California but returned in 1880 to his friends on the Mississippi where he spent the rest of his life piloting and swaping stories. He was a half-brother of William Irvine.

In 1862, Knapp, Stout & Company disposed of *Maquoketa City* for $3,250, a boat they paid $2500 for the year before. The outbreak of war caused prices to rise.

The *Frank Steele*, a small side wheeler made her appearance on the Chippewa in 1862. She was owned by Wm. F. Davidson of La Crosse and was chartered by the federal government for war duty.

In April, 1862, the *Pomeroy*, a Mississippi tramp, made her appearance at Eau Claire, but never paid a second visit. The *Flora Temple* also came up the river that year and was sold to H.T. Rumsey who was accumulating a small fleet of steamers for the river trade.

The summer of 1862 also saw the arrival of *Monitor*, a geared side wheeler of ten tons "the lightest and fastest boat on the river" (built) expressly for the passenger trade," according to an advertisement in the *Free Press* dated June 17, 1867. It ran mainly between Durand and Read's Landing to connect with a stage line out of Wabasha. The boat was actually built at Read's by John C. Thorpe of Eau Claire and Benjamin E. Seavy (also as Seavey). Captain Seavy was the first to build the geared side wheeler with twin engines. One of his boats was sold to H.T. Rumsey who named her *Johnny Schmoker* no doubt to honor a friend named Abraham Schmoker, a Swiss pioneer who settled on a farm near Beef Slough and cut wood for the steamboat trade. (All boats burned cordwood in the firebox. A photograph of the *Phil Scheckel* tied alongside shows a pile of cordwood on the levee, cut to standard length of four feet.)

The late Mildred Houghton Comfort, formerly of Winona, Minnesota, wrote two novels based on the *Schmoker*, one called *Winter on the Johnny Smoker*, and the other *The Treasure on the Johnny Smoker.* She altered the spelling slightly from the original. Her books went into several editions.

In 1859, according to Merrick, the *John Rumsey* appeared on the Chippewa River. She was a stern wheeler of forty tons built from parts taken off the *Flora Temple.* Using parts from one boat to build another, or exchanging boilers, seemed to be a constant source of employment on the river. Boats were lengthened and their engines enlarged. They were all custom made and some did not handle as well as others when first launched and changes had to be made.

In 1862, the *Golden Star*, a small geared side wheeler was built by Phil Scheckel and William Warran and sold to Captain E.E. Heerman, one of the most widely known captains on the Mississippi or Chippewa rivers.

The *Willie Spray*, a side wheeler, made a few trips up the Chippewa River in 1863.

The *Idell Prindle*, a stern wheeler, was built by Miles Prindle in the winter of 1863-64, the same winter that Knapp, Stout & Company built another geared side wheeler called the *Pete Wilson* which made her first run in the spring of 1865. Phil Scheckel was her pilot for several years.

Steamboats first seen in 1866 include the *Phil Sheridan* and the *Netta Munn*, the latter built at Durand by J.A. Cavanough and M.A. Lewis. The *Phil Sheridan*, however, was not seen on the Chippewa River long and ran mostly between St. Paul and St. Louis.

The *L.W. Barden*, a side wheeler of 102 tons, built on the Wolf River in eastern Wisconsin, was used by the Daniel Shaw Lumber Company in 1866 to push lumber rafts from Read's down the Mississippi. Walter A. Blair, in his *A Raft Pilot's Log*, says the crew called her the "L.W. Workhouse." (See accompanying photo.)

Other boats seen on the Chippewa River in 1866 were the *Julia Hadley*, *Champion*, *Cutter*, and *Adventure.* Blair says *Julia Hadley* and *Champion* were among several boats which "wore themselves out in the work" and were never rebuilt. The *Julia Hadley*, by 1867, was running between Eau Claire and La Crosse three times a week.

The *Silas Wright* appeared on the Chippewa in 1867. She was built at Durand for the Chippewa Express Company, and in 1869, according to Blair, she was acquired by Porter & Moon (later Northwestern Lumber Company of Eau Claire). Captain J.M. Turner was her master for several seasons. An advertisement in the *Free Press* on April 11, 1867, boasted that this "ship was the only reliable Chippewa boat."

The *Ida Campbell* already mentioned, was built at Durand and named after a daughter of Ed Campbell. Some time after 1871 Campbell was forced to sell his boat and he moved to Long Lake to operate a stopping place on the Chippewa Trail.

Most of the time the Chippewa River was too risky for steamboat captains to indulge in racing, a common event on the Mississippi, but in 1866, the *Monitor* and *Johnny Schmoker* engaged in a race to see which was best. According to the Wabasha *Herald* of November 16, 1866, *Johnny Schmoker* won the race which gave her the honor of mounting a broom on the jackstaff where the flag usually flew. The following spring, when *Johnny Schmoker* came up the Mississippi River, *Monitor* was waiting at Read's Landing, spoiling for another race. But this was no bow-to-bow affair, with smoke-chimneys belching black smoke. *Monitor* left Read's Landing an hour ahead of *Johnny Schmoker* and reached Eau Claire two and a half hours ahead of her after four stops. *Johnny Schmoker* allegedly made eleven stops. It is difficult to understand how the timing on this was computed in view of the difference in the number of stops. Both boats claimed the honor of being first and that ended the competition.

The *Annie Girdon*, forty-eight tons, owned by Knapp, Stout & Company, came into Dunnville in 1867 to push company lumber rafts down to Read's Landing. Here, after coupling up with more Chippewa rafts, the larger "Mississippi raft" was pushed downstream. The *Annie Girdon* and *Union* were probably the first to push rafts on the Mississippi River.

Up to June 20, 1867, a total of 165 steamboats

The *Phil Scheckel* (upper left) was a side wheeler when it was built in 1878. Henry Carlisle, (in white sleeves) was pilot under Captain Scheckel. Lower photo shows the *Phil Scheckel* after she was lengthened in 1896 and made into stern wheeler. Inset left: Phil Scheckel, and inset right, Ida Scheckel, his wife. Upper right, the remodelled boat pushes a lumber raft down Mississippi River.

The *Red Wing*, 150 tons, built in 1854, was one of largest on upper Mississippi and for several years in late 1850s dominated traffic between Eau Claire and La Crosse. She was commanded in early years by Captain John Woodbarn. Year 1870 painted on picture probably marks change of ownership to "Northern Line Packet." Picture taken, no doubt, for new owners. Crewmen are lined up on roof deck. There is no wind blowing dust off levee in foreground, yet flags are straight out, suggesting picture was touched up by the photographer.

The *Champion* was first seen on Chippewa River in 1866 and featured ornamental pilot house. Note ship's bell at left of man seated. Railroad track at right suggests picture taken on Minnesota side of Mississippi.

had stopped at the Eau Claire levee on the West Side. On July 9th, three docked on one day, the *Silas Wright*, *Ida Campbell* and *Johnny Schmoker.* These boats were carrying mostly freight.

In the spring of 1868, one of the new boats to enter the Chippewa River trade was the *Mollie Mohler*, 121 feet long, which ran daily for a time between Eau Claire and Read's Landing.

These steamboats were always at the mercy of a relentless river; the stage of water was either too high, or too low. And some pilots pressed their luck too late into the season. On November 25, 1869, the *Free Press* reported that *Ida Campbell* had arrived at the levee after a "hopeless voyage of eight days from Read's Landing, combatting ice, snow and low water." At Plum Island, below Durand, she sank a barge loaded with freight for Eau Claire merchants but most of the cargo was recovered. The *Johnny Schmoker* was also pressing her luck and tried to get up the river but had to turn back.

Captain Heerman thought he could still make it to Durand on the *Union*, and on December 16th, he got as far as Marks Landing where the *Union* broke some cogs in her core wheel. Mark's Landing was a stopping place for raftsmen returning on foot from Read's to Eau Claire or Menomonie, and it was here that George Ecklor built a hotel in the 1860s to cater to this traffic. Someone from Read's Landing ran a store here and called the place "Shoo Fly" after a song of the period, and soon the entire community was being referred to as Shoo Fly, a fact which caused no end of jokes and merriment. But when a post office was established there in 1869, it was called Ella, and though the post office closed in 1904, the community still calls itself Ella.

In 1869, Ed Campbell came up the Chippewa River on his new boat, the *Frank Forest*, Captain Chet Hall in command, Eli Miner, pilot, and George T. Davis, engineer. Her boiler was new but her machinery came out of another boat, the *Jeannette Roberts.* For her coming out party, an excursion was run on September 14th from Eau Claire and Durand to St. Paul. Forty couples joined the cruise.

Captain Heerman tried out his new stern wheeler, the *Minnietta*, in November, 1869, the season just about over, but he wanted to make sure that everything was in order for the 1870 season.

The *Phil Sheridan* did not have steady trade on Chippewa River but ran up on demand. She was owned at this time by Northwestern Union Packet Co. A mural, presumably commemorating Phil Sheridan's famous ride at Winchester, adorns wheel house. Charles Edward Russell refers to *Phil Sheridan* as the "pride and wonder of the river, the swiftest thing that ever floated there . . ." After outlasting most steamboats, the *Phil Sheridan* smashed up in crash of ways while undergoing repairs at La Crosse.

Ed Campbell ended his steamboat career on the Chippewa when the *Ida Campbell* was confiscated by the sheriff's department for debt. People said he

spent too much money building his new boat, the *Frank Forest.*

In 1870, *Buckeye*, by now a well-known Mississippi River boat engaged in rafting lumber and logs, made her last run to Chippewa Falls. The *Clyde*, built with a steel-plated hull, the first of its kind, was shown to her new owners, Ingram & Kennedy, at Eau Claire and was sent back to the Mississippi to push company rafts. The steel covered hull was supposed to protect it against underwater hazards and stray logs.

In early January, 1871, the *Ida Campbell* burned. The *Johnny Schmoker* and *Chippewa* burned later in the month while tied up at Rumsey's Landing below Eau Claire. Incendiarism was suspected in the burning of the *Ida Campbell* and *Johnny Schmoker*, although the latter was not seriously damaged. Captain Heerman bought the wreck of *Ida Campbell* and brought her to Dead Lake to canabalize her for a new *Ida Campbell* which was ready for her maiden voyage by spring. Heerman built or rebuilt several boats, some of the same name.

In the years that followed, that is, in the 1870s, one boat after the other dropped out of the trade. Competition from the railroads was forcing boat owners to find new ways or new rivers to make a living. Captain Heerman sent one of his boats up the Missouri to Fort Benton, and another he dismantled and shipped by train from St. Paul to Larrimore, North Dakota, the end of the line. Here wagons were used to haul the parts to Devil's Lake where the boat was assembled and used as a laker.

In 1877, only two new boats ascended the Chippewa, one, the *Silver Star*, a stern wheeler of about thirty tons, built at Chippewa Falls and commanded by Captain William M. ("Noisy") Smith, who seldom had anything to say and when he did, he was scarcely audible. The other new boat on the river that year was the *Bruno* built in Illinois in 1874 which Knapp, Stout & Company acquired through a foreclosure. She was taken to the company's boat works at Waubeek for overhaul but was found beyond repair.

Phil Scheckel, master of the *Phil Scheckel*, married Ida White, also an immigrant from Austria whom he met in this country. They made their home at Waubeek. They are buried at Eau Galle.

Other steamboats sailing under the Knapp, Stout flag in the 1870s and 1880s were the *Helen Mar*, W.R. Slocomb, master, the *Louisville*, Andrew Larkin, the *Menomonie*, Stephen Withrow, and the *Bart E. Linehan*, Lafe Parket. By 1893 most of these rafters had been sold to the La Crosse Mississippi Towing Company.

Knapp, Stout & Company of Menomonie probably built or commissioned more steamboats than any other lumber company in the Chippewa Valley. The first one to be used as a "rafter," that is, to push lumber rafts, was the *Annie Girdon* which shared honors with the *Buckeye* as one of the first steamboats to successfully push a lumber raft down the Mississippi River.

In 1878 the company built the *Phil Scheckel*, a small side wheeler, which, in 1896, was lengthened into a powerful stern wheeler. Her master, Phil Scheckel, was a native of Austria who had immigrated to the United States in 1855. He later worked as an employee of Knapp, Stout & Company and finally achieved his life's ambition to become the master of a boat named after him. He pushed the last lumber raft down the Chippewa from Dunville in 1901, at which time the company sold the boat to the Florida East Coast Railway where she was used as a tender in building the highway from Miami to Key West. In 1919 the durable *Phil Scheckel* was sold for parts.

Steamboating was made largely obsolete when the railroad came to Eau Claire in 1870 because freight could be shipped cheaper on the "cars" than by boat. *Ida Campbell* hung on for a while and in 1881 was advertising passenger tickets from Eau Claire to Read's Landing for $3.50, and $1.50 to Durand. After 1870, according to J.N. Macomb, Corps of Engineers, no steamboats reached Chippewa Falls.

Other boats which ascended the Chippewa at one time or another would include the *Minnie Herman*, *Alvira*, *Annie May*, *Alone*, *Albany*, *Jessie Bill*, *R.F. Weaver*, *Johnnie Hawkins*, *Dr. Franklin II*, *Eau Claire*, *Hudson*, *Iowa City*, *Silver Lake*, *Wm. Hyde Clark* and *Artemus Lamb*.

The *Jessie Bill*, according to Blair, was doing "all kinds of company work" for Beef Slough Company.

Another steamboat, the *Alice*, probably came up the Chippewa River at one time or another, but information on this ship is contradictory and it has not been determined whether she was built in 1853 or in 1863-64. A ship of seventy-two tons, built in Pennsylvania, she reached St. Paul, according to Merrick, in 1854.

In the several decades of steamboating on the Chippewa River, although there were men who drowned while struggling to free a steamboat from a sandbar, or by other accidents, there never was a fatality to crew or passenger which could be attributed to the pilot or master of the steamboat—a high tribute to the competence of the men who once navigated the Chippewa River.

The *Silas Wright* was built at Durand for Chippewa Express Company, probably 1866, and ran every season until September 3, 1892, when, acting as bowboat on big lumber raft, she was crushed in accident at Le Claire Rapids (above Rock Island) on Mississippi. Porter & Moon (later Northwestern Lumber Company) was her owner for a time. Door at right on roof deck is near galley; dustpan, washpan and broom hang outside. Photo probably taken at Read's Landing.

Names in ballad below are steamboats which once ran Mississippi, St. Croix and Chippewa rivers, most of them rafters, but author unknown. Poem was kept alive by oral tradition and finally reached print in Burlington (Iowa) *Post* in 1927, and copied by Charles Edward Russell in book published in 1928 called *A-Rafting on the Mississippi.* Accompanying photos show *Anna Girdon* was spelled *Annie Girdon* and *Menomonee* was spelled *Menomonie.* The *Mollie Mohlaer* should probably be spelled *Mollie Mohler.*

MISSISSIPPI STEAMBOAT BALLAD

The *Fred Weyerhauser* and the *Frontenac,*
The *F.C.A. Denkman* and the *Bella Mac.*
The *Menomenee* and *Louisville,*
The *R.J. Wheeler* and *Jessie Bill.*
The *Robert Semple* and the *Golden Gate,*
The *C.J. Caffery* and the *Sucker State.*
The *Charlotte Boeckler* and the *Silver Wave,*
The *John H. Douglas* and the *J.K. Graves.*
The *Isaac Staples* and the *Helen Mar,*
The *Henrietta* and the *North Star.*
The *David Bronson* and the *Nettie Durant,*
The *Kit Carson* and the *J.W. Van Sant.*

The *Chauncey Lamb* and the *Evansville,*
The *Blue Ridge* and the *Minnie Will.*
The *Saturn* and the *Satellite,*
The *LeClair Belle* and the *Silas Wright.*
The *Artemus Lamb* and the *Pauline,*
The *Douglas Boordman* and the *Kate Keen.*

The *I.E. Staples* and the *Mark Bradley,*
The *J.G. Chapman* and the *Julia Hadley.*
The *Mollie Whitmore* and the *C.K. Peck,*
The *Robert Dodds* and the *Borealis Rex.*
The *Pete Kerns* and the *Wild Boy,*
The *Lily Turner* and the *St. Croix.*

The *A.T. Jenks* and the *Bart Linehan,*
The *C.W. Cowles* and the *Brother Jonathan.*
The *Pete Wilson* and the *Anna Girdon,*
The *Inverness* and the *L. W. Barden.*

The *Nellie Thomas* and the *Enterprise,*
The *Park Painter* and the *Hiram Price.*

The *Dan Hines* and the *City of Winona,*
The *Helen Schulenberg* and the *Natrona.*
The *Flying Eagle* and the *Moline,*
The *E. Rutledge* and the *Josephine.*
The *Taber* and the *Irene D.,*
The *D.A. McDonald* and the *Jessie B.*

The *Gardie Eastman* and the *Vernie Swain,*
The *James Malbon* and the *L.W. Crane.*
The *Sam Atlee* and *William White,*
The *Lumberman* and the *Penn Wright.*
The *Stillwater* and the *Volunteer,*
The *James Fisk Jr.*, and the *Reindeer*
The *Thistle* and the *Mountain Belle,*
The *Little Eagle* and the *Gazelle.*

The *Mollie Mohlaer* and the *James Means,*
The *Silver Crescent* and the *Muscatine.*
The *Jim Watson* and the *Last Chance,*
The *Kate Waters* and the *Ed. Durant.*
The *Dan Thayer* and the *Flora Clark,*
The *Robert Ross* and the *J.G. Park.*
The *Eclipse* and the *J.W. Mills,*
The *J.S. Keator* and the *J.J. Hill.*
The *Lady Grace* and the *Abner Gile,*
The *Johnny Schmoker* and the *George Lysle.*
The *Lafayette Lamb* and the *Clyde*
The *B. Hershey* and the *Time* and *Tide.*

The *L. W. Barden* was built on Wolf River and acquired by Daniel Shaw Lumber Company of Eau Claire to use as a "rafter," that is, to push lumber rafts on Mississippi River. She made several runs up to Chippewa Falls to pick up cargo of leather products made in Cadott and hauled from tannery to river.

The *Menomonie*, another stern wheeler in service of Knapp, Stout & Company as a rafter. Company logo appears on starboard hull, only steamboat apparently to show K (for Knapp) within S (for Stout).

Oral tradition along the Chippewa River holds that this steamboat, *Luella II* came up the river in 1880s on a campaign cruise to drum up interest in Populist movement of Jacob S. Coxey, later the leader of Coxey's army in the march on Washington in 1894. *Luella II* was owned at one time by the C. Jellison Towing Company of Wabasha. Men and women on roof deck of steamboat are well dressed, and one holds up "Coxey" poster. Cordwood on lower deck was used for firebox, and building in background at left seems to be "Machine Shop."

The *Louisville*,, built for Knapp, Stout & Company, was used as a rafter in the 1880s, pushing lumber rafts from Dunville, on Chippewa River to Read's Landing on Mississippi where Chippewa rafts were assembled into large Mississippi rafts and pushed downstream to lumber retail yards in river towns all the way to St. Louis, and some as far as New Orleans.

Weyerhaeuser's Ditch

The shortest route between Durand in Pepin County, Wisconsin, and Nelson in Buffalo County is Highway 25. South from Durand in only a matter of minutes the highway, clinging to the high bluffs on the left, swings left and there, in plain sight, is a hill which dominates the surrounding terrain. It is not a big hill, but it stands out sharply because the plain below is part of the Chippewa River "bottoms" or flatlands through which the river flows south to its mouth on the Mississippi River.

Since the hill seems to be round at the bottom and rises sharply to a mound it has, since early times, been called "Round Hill." It was a landmark to the Indians and to the early French explorers who camped at its base, and it was a beacon to the lumber raftsmen and log drivers who came this way in the heyday of the logging era from 1850 to 1905. The hill, apparently, has always been covered with a growth of trees of little commercial value and there is no evidence that the wood was ever used for anything but firewood by succeeding generations of farmers.

If anyone had been standing at the top of Round Hill in the year 1880 he or she would have seen an extraordinary sight, for to the south lay a long log boom resembling a huge serpent dangling obliquely across the Chippewa River from the west bank just below the hill. This boom was stretched across the river to sheer, or bounce, incoming logs away from the main channel of the Chippewa River into a slough known since early times as "Beef Slough."

Although called a slough, Beef Slough was more like a slow-moving river, but important to lumbermen because it could be controlled without the expense of maintaining a dam upon it. Below Nelson and extending east to Alma along the lower course of the slough, there were several small lakes which were ideal for storing logs, sorting them, and coupling them into "brails" or multiple log rafts which could be floated down the Mississippi River, or pushed down by steamboats.

A single log boom is constructed of two or three squared logs lying abreast, held together by chains or spikes, and covered on top by planking wide enough to make a walk. Single booms were linked to one another, end to end, by link chains for the distance required to sheer or bounce incoming logs on the river in one direction or the other.

One boom below Round Hill was said to be more than 2000 feet long, but it did not extend all the way to the mouth of Beef Slough. For one thing, it was a fixed or rigid boom, anchored at the north end by a heavy chain ashore and on the south end to pilings in the river. Attached to the same pilings below the fixed boom was a fin boom, that is, a boom equipped with rudders or fins on the lee side.

The fin boom, the invention of two Eau Claire lumbermen, was built with planks or fins dangling from the lee side. On some fin booms a line, attached to the ends of the fins, ran back to a windlass anchored ashore or to a pier where the boom boss could control the action of the fins. If he wanted to open the boom to allow a steamboat to pass through, he wound the crank on the winch to pull the fins in, or parallel to the boom, and the current pushed the boom toward shore. After the steamboat passed downstream, the boom boss loosened the line on the fins and they swung outward on the back, or lee side of the boom. The boom now floated back to midstream, since the volume of water striking the fins was greater than the volume of water striking the leading edge of the boom.

The fin boom below Round Hill, however, was not controlled by a line running back to a winch. According to William H. Gilmore, whenever it was necessary to manipulate the fins, crewmen at Round Hill walked down the fixed boom onto the fin boom and, with their peavies or pike poles, manipulated the fins manually.

Gerhard Gesell took a picture of a fin boom anchored above Alma, probably in 1889, and there is reason to believe that it is the Round Hill fin boom which had been towed down the Chippewa into the Mississippi after operations of the Beef Slough Company closed down on the Wisconsin side and moved across the Mississippi to West Newton on the Minnesota side.

The fin boom below the pilings at Round Hill bounced logs into the slough, but it could, as explained above, be swung in or out to allow passing steamboats or lumber rafts through.

Round Hill as seen from bluffs on Highway 25 in Buffalo County. Chippewa River winds to left of hill and continues south, about twenty miles, to Mississippi River.

Both booms below Round Hill were augmented by a wing dam which extended north from Horse Island near the mouth of Beef Slough. This was made of rip-rapping and brush, but even with this wing dam reaching toward the fin boom, there was a gap between the south end of the fin boom and the north end of the wing dam where logs slid through. In order to catch the strays, another wing dam was built below Ecklor House. This shunted the strays into "Little Beef" river and thence into the main slough.

Into the mouth of Beef Slough went tens of thousands of pine logs each year from trees cut in the Chippewa Valley. They had floated, twisted, turned and thumped each other all the way down the Chippewa River and then bounced off Round Hill boom into the slough to be sorted and made into brails for steamboats to push down the Mississippi to sawmills all the way from Winona to St. Louis.

On January 30, 1875, Major F.A. Farquhar, resident engineer on the upper Mississippi River, sent a report to his superior, Brig-Gen. A.A. Humphreys, chief of U.S. Corps of Engineers, on the status of the Chippewa River with special reference to the boom works below Round Hill. General Humphreys sent the report on to the Secretary of War and this, in part, is what the major had to say:

> "At Round Hill the Beef Slough Company have constructed a line of boom piers to support a sheer-boom that diverts all logs to the left bank of the river and leaves only a passage between the end of the boom and the rocky shore of 130 feet.
>
> Just above the head of Beef Slough is another sheer boom to divert all logs to the left channel; and finally this left channel is entirely closed by a line of boom piers constructed with a sheer boom. The effects of the above booms are to make it very difficult for rafts coming down stream to pass to the right of Beef Slough Island, and to deflect an undue amount of water into Beef Slough, which water is here much wanted in the main river below.
>
> I am informed that the boom company is enjoined by its charter not to leave obstructions to navigation in the river, and not to deflect any water from the main stream not already (at the time of its incorporation) flowing in Beef Slough.

When this was written, it seems that there were two sheer booms below Round Hill, one extending

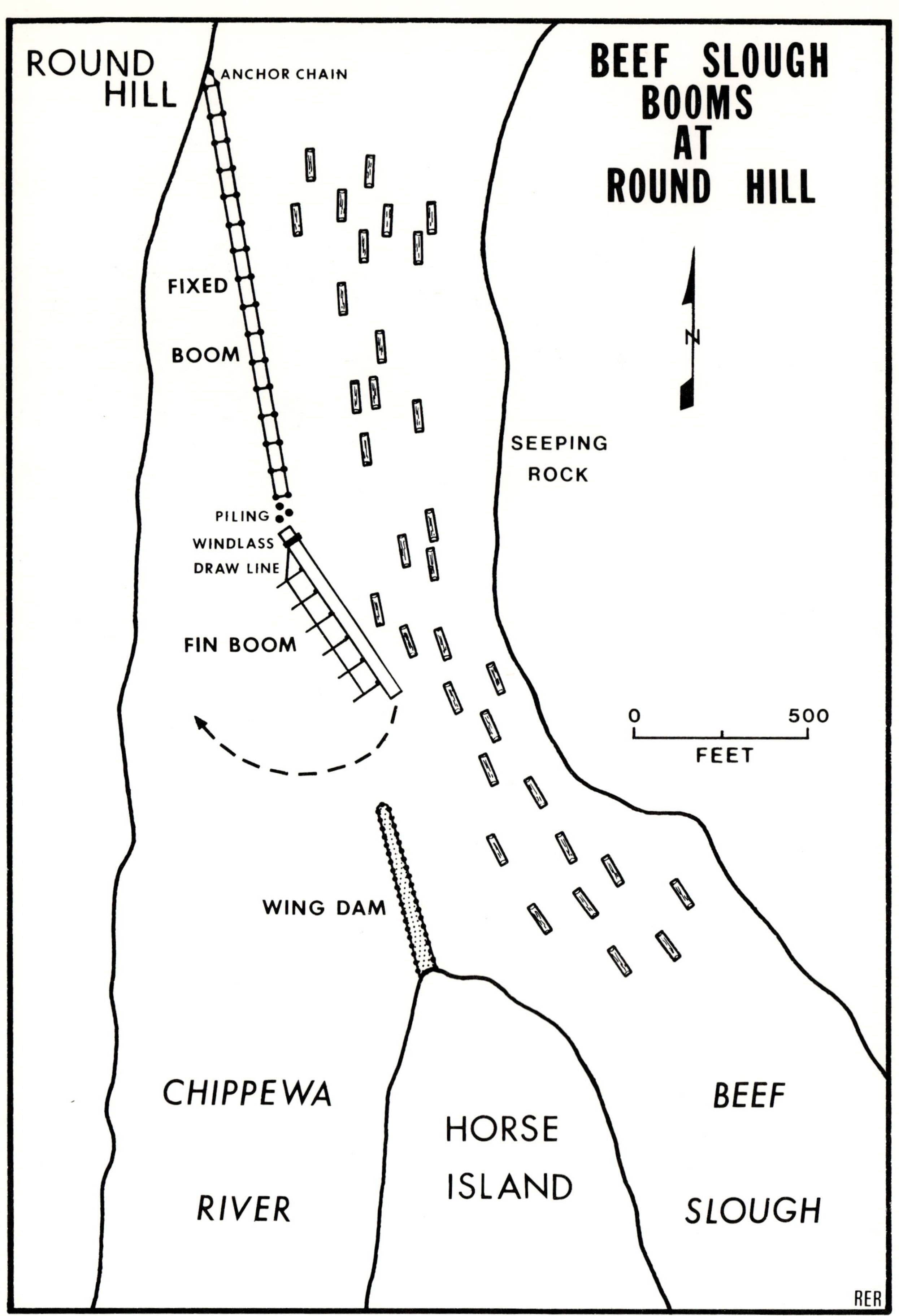
BEEF SLOUGH BOOMS AT ROUND HILL
ROUND HILL
ANCHOR CHAIN
FIXED
BOOM
SEEPING
ROCK
N
PILING
WINDLASS
DRAW LINE
FIN BOOM
0
500
FEET
WING DAM
CHIPPEWA
RIVER
HORSE
ISLAND
BEEF
SLOUGH
RER

downstream from the anchor chain in front of "Ab" Gilmore's house, and still another closer to the entrance to Beef Slough, leaving a gap between them of 130 feet for passage of steamboats and lumber cribs.

The accompanying map drawn by Randall E. Rohe, based on testimony of William Gilmore, shows that the arrangement described by Major Farquhar had been altered a bit, for instead of a boom running down to the entrance of Beef Slough, a wing dam extended upstream from Horse Island. But there were still two sheer booms above the wing dam. Also, the passage for steamboats now lay between the lower end of the lower sheer boom and the upper end of the wing dam.

It seems a puzzle, though, why Major Farquhar had to be "informed" that the boom company had been enjoined by its charter not to leave obstructions to navigation in the river, and not to deflect any water from the main channel of the Chippewa River into Beef Slough. It was common knowledge that the stage of the Chippewa River below Round Hill had been lowered by the booms below Round Hill, and the stage of water in Beef Slough raised.

But if the original stockholders of the Beef Slough Company had been "enjoined" by their charter not to obstruct the river and not to deflect any water into Beef Slough, it also appears that Weyerhaeuser, who inherited this charter, either did not read it or ignored it, because obstructions to navigation below Round Hill continued and water from the Chippewa River was diverted to Beef Slough.

Finally, in 1879, Captain E.E. Heerman, one of the senior pilots on the Chippewa, brought suit against the Weyerhaeuser companies for damages to one of his boats caused by loose floating logs.

Map by Randall E. Rohe of Beef Slough booms below Round Hill on Chippewa River. First boom built in late 1860s extended south all the way from Round Hill to the entrance to Beef Slough with fin boom in center for passing steamboats and lumber rafts. Map shown here represents changes made, probably about 1880. This arrangement probably continued to be used until Beef Slough was abandoned in 1889.

Weyerhaeuser allegedly muttered to associates that Heerman should pay him for damaging his logs. In court, it was generally understood that Heerman was not acting entirely on his own but had the backing of Eau Claire mill owners. Thomas Irvine, secretary for M.R.L. Co., who resided much of the time at the company house below Nelson's Landing, was awakened in the middle of the night of January 31, 1879, and served with a summons in the suit of Heerman v. Beef Slough . . . *et al.* Heerman was asking for an injunction to restrain the Weyerhaeuser people from running loose logs on the Chippewa, and he asked the court for an order to have all booms, dams and piers in Beef Slough removed.

The court ruled nearly a year later that the obstructions were not "to be abated by a court of equity, even if the plaintiff had shown that he was especially damaged by them, which he has not." It was also the court's opinion that the logging interests who were using the Chippewa to move their logs were "very large" and that the alleged interference with navigation had been "acquiesced in for a long time."

And that settled the matter of steamboat v. logs. Both were entitled to use the river, and, since logs outnumbered steamboats half a million to one, there was no contest. Lumbering formed the basis of almost the entire economy of the Chippewa Valley. Pine Tree was King and Weyerhaeuser was Boom Boss.

There is nothing left today of Beef Slough and its extensive works, no pocket booms, no sorting gaps, no pilings—nothing. Even the course of the slough has changed in the past century and where it was once a wide stream, it is now a trickle, and where there was once a lake, there is a marsh. To cap it all, most of the slough below Nelson is flooded by U.S. lock and dam No. 4 on the Mississippi River at Alma. A heavy piece of chain, the last vestige of the long boom below Round Hill, dangled in the Chippewa River until a few years ago when a souvenir hunter came up the river, presumably in an outboard, and cut off part of the chain with an acetylene torch.

When Weyerhaeuser abandoned the slough in 1889, the entrance to it, already badly silted, was further blocked off by the simple expedient of sinking several cribs of green lumber into the mud. William Gilmore, who lives a short distance upstream on the site of the old camp where his grandfather lived, believes the lumber is still there, probably well preserved. The wing dam which extended from Horse Island was blasted in the center with dynamite to open a new channel for the Chippewa River, away from the old mouth to Beef Slough. The entire wing dam has since disappeared but in low water it can be traced.

Why was it necessary to use a slough, or more accurately, an auxiliary river to float logs into the Mississippi? Why not go straight down the main channel of the Chippewa below Round Hill and run the logs into the Mississippi? But it was this very straight shoot that made it unsuitable for a log pond and sorting works. This would have required a dam on the Chippewa this close to the Mississippi and it is doubtful whether anyone ever gave that a serious thought.

John R. Fonda, in his "Reminiscences of Wisconsin," written for the *Wisconsin Historical Collections* (1867), remembered Beef Slough as a "pretty stream of water, wide and deep, with fine banks." He was recalling an incident of nearly forty years earlier in 1829 when he, together with a detail of troops from Fort Crawford, was sent into the Pinery to bring back some square timbers and shingles needed for the construction of Fort Crawford at Prairie du Chien, Colonel Zachary Taylor commanding.

The soldiers got their square timbers, probably from the first sawmill built on Wilson Creek on the Red Cedar River, and they made a raft to run the timbers down the river into the Mississippi. All

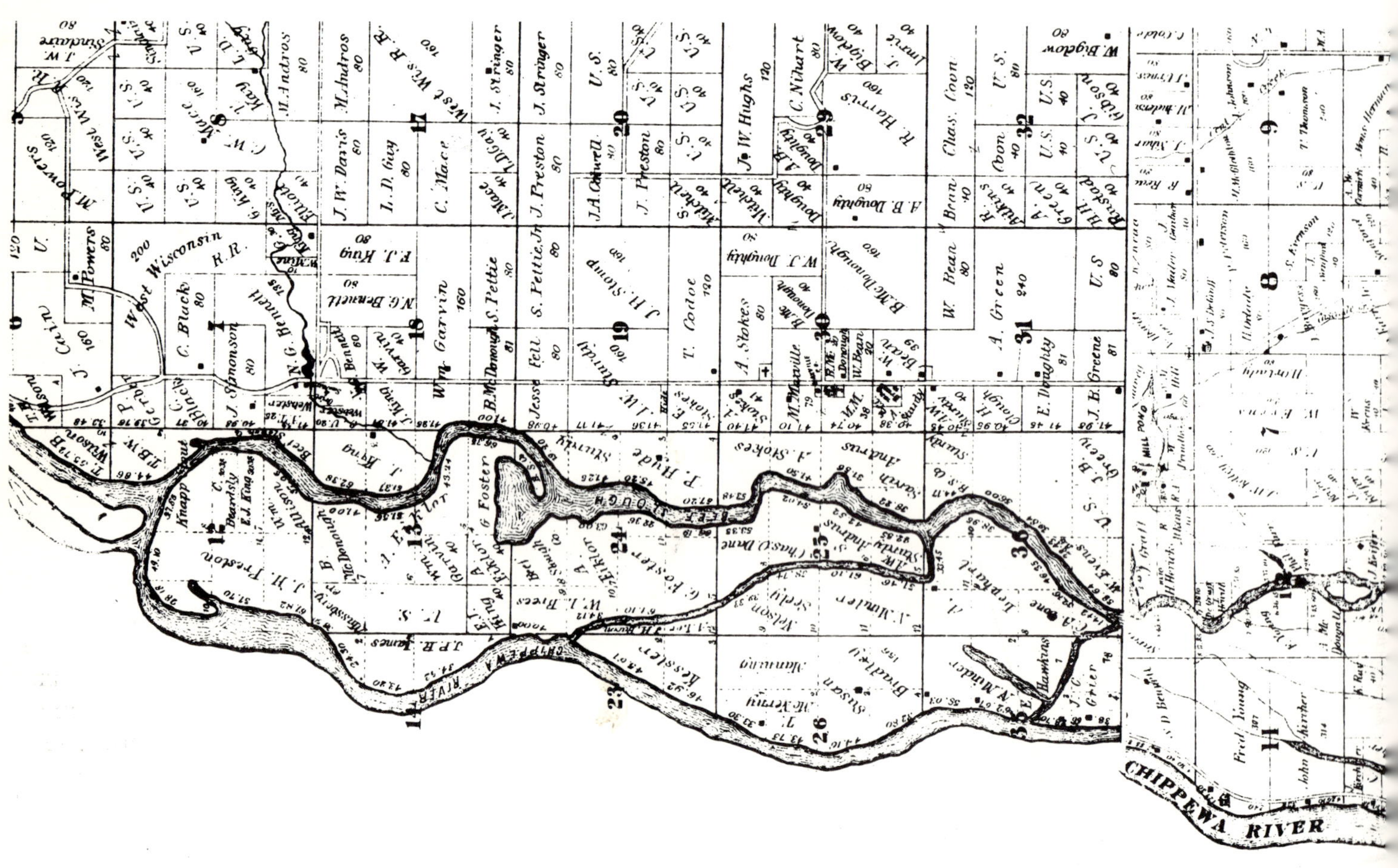

went well until the raftsmen reached the entrance to Beef Slough and, thinking this was a short cut to the Mississippi, they boldly ran their rafts into it and struck a wall of driftwood which smashed the raft. After much work, the pieces were collected and a new raft was made, but no one tried the slough again until three decades later. By that time, lumbermen on the Chippewa had come to see that the "pretty stream of water" touted by Fonda would need considerable improvement before it could be used for the movement of logs.

The origin of the name "Beef Slough" often stirs controversy. Modern highway maps and plat maps contradict each other on the names of rivers in Buffalo County, and the Buffalo River, which empties into the Mississippi north of Alma, is sometimes called "Beef River" and Beef Slough is seen on some maps as "Buffalo Slough." Perhaps the best explanation for the name comes from the pen of Reuben Gold Thwaites who, in the *Wisconsin Historical Collections* (1908), has this to say:

> The river Bon Secours has been correctly identified with the present Chippewa River of Wisconsin. But it should also be noted that it is the same stream as that called by the early explorers "River des Boeufs," now known as Beef River (Wis). The mouth of the Chippewa has shifted since the seventeenth century . . . when it entered the Mississippi at the southern end of what is now called Beef Slough. This would make Beef River but a tributary of the Chippewa . . . On some of the old maps the names are given as alternates, 'Bon Secours on Beef River.' LeSueur in 1700 . . . gives the reason for its name (Good Help) in the large number of buffalo and game found thereon.

Meanwhile, there were rumors in the 1860s that

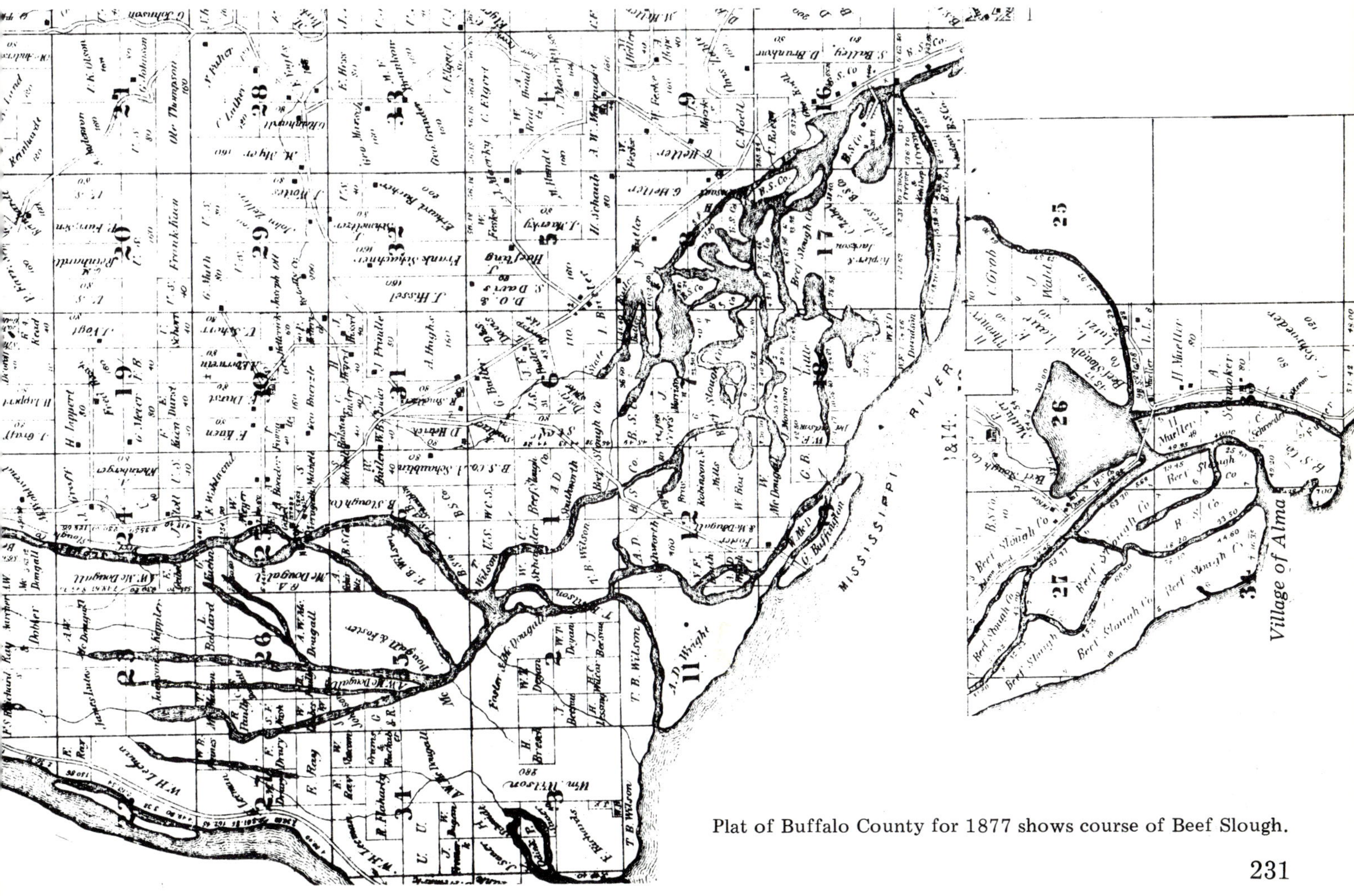

Plat of Buffalo County for 1877 shows course of Beef Slough.

Street scene in Durand, 1876. Rear of livery stable faces levee of Chippewa River. Tall man at right is William P. Hungtingon, former riverboat clerk who settled at Durand to publish first newspaper. He also operated livery stable. In center, August Rielle, blacksmith. Holding hand of Hungtinton is daughter, Amelia, who later married William A. Gilmore. Holding horse at left is Henry A. Knapp. Slogan after his name reads "A little higher." He was buying wheat.

Mississippi River mill owners were going to organize a company to exploit the advantages offered by Beef Slough for log storage and sorting. One of the reasons they wanted this slough was blamed on the delay which the Mississippi River people often encountered in getting their logs down the Chippewa River. Matthew G. Norton, not entirely an unbiased witness, writes in his book on the *Mississippi River Logging Company* (1912) that the Chippewa Falls mill owners "held in their booms the logs which should come down the river to the Beef Slough Company, and sought in every way to obstruct the progress of the Company." This jumps slightly ahead of the story, but the same delaying tactics were charged by Eau Claire mill owners against Chippewa Falls even before Beef Slough Company was organized in 1867.

When rumors floated up the river to the Menomonie and Eau Claire mill owners that the Mississippi River people were looking at Beef Slough, the upriver millmen got a franchise from the state legislature allowing them to string a boom into Beef Slough. They had no intention of using it; they wanted to prevent anyone else from using it. The plat of 1877 of Buffalo County reveals that three or four forties of land around the head of Beef Slough were still owned by Knapp, Stout and Company, and T.B. Wilson, an officer of the company. But no one could halt a log on any river or slough even if the river ran between two forties of another man's land, and the Mississippi River mill owners, all the way from Winona to Rock

Island, wanted access to the pine in the Chippewa Valley and they were determined to get it.

On August 13, 1867, a consortium of mill owners from Michigan and Wisconsin was incorporated at Alma to form the "Beef Slough Manufacturing, Boom, Log Driving and Transportation Company," abbreviated to "Beef Slough Company," for the purpose of "driving and booming logs, manufacturing lumber, doors, etc. and of using Beef Slough for log driving and rafting." One of its first tasks was to acquire land along the slough below the Knapp, Stout forties, and the plat of 1877 reveals that the company, at least by that time, owned nearly everything along the slough in the town of Nelson, Buffalo County.

The purpose of the new company ran counter to a franchise obtained by the Menomonie and Eau Claire mill owners, and preliminary efforts by the new organization to get a franchise from the state legislature of Wisconsin passed the assembly but was rejected in the senate, probably as a result of political pressure from the former group.

The Mississippi River mill owners found another way. The 1869 legislature enacted Chapter 331, Private & Local Laws, incorporating the Portage City and Light Company, Section 9 of which inconspicuously enacted a piece of general legislation providing that "in all cases where any franchise or privilege has been or shall be granted by law to several persons, the grant shall be deemed several as well as joint, so that one or more may accept and exercise the franchise as though the same were granted to him or them alone."

One of the men involved, James H. Bacon, now switched sides and claimed the right to act individually under Section 9, and assigned his right over to the Beef Slough Company. Law makers and mill owners from Eau Claire and Menomonie had been caught napping and someone had managed to slip into the bill related to Portage City a seemingly innocent amendment which no one had noticed.

Yet, the position of the company was tenuous. No one seemed to be sure who controlled the slough, but the company went ahead, as if it did, and built improvements costing $11,000. Not long after, another $5,000 was allotted for further improvements. By 1889, when Beef Slough was abandoned, about half a million dollars had been expended in booms, piers, pilings, pocket booms, coupling works, camps for crews, offices for local managers, clerks and scalers, and barns for horses and cattle. At the height of season in mid-summer every year there were about 1,500 men employed at one job or another all the way from Round Hill to the coupling works below Nelson.

In addition to the costs of improvements, the company faced another hurdle. In 1868, of the fifty million feet of stumpage contracted for in the Chippewa Valley, only ten million feet reached Beef Slough. Nature was partly to blame. Many of the logs were delayed by jams, as well as in booms on the Chippewa and its tributaries, and many logs were stolen outright by mill owners and farmers along the way. As late as June, 1868, Beef Slough logs were still being held up, especially in Eau Claire booms.

Ecklor House on Dead Lake on north side of Ella where raftsmen, returning on foot from Read's Landing, stopped over night en route to Menomonie or Eau Claire. Steamboat captains also allowed passengers to go ashore here to enjoy hospitality and fine food of Ecklor family. Business was hurt by flood of 1880 which created big sandbar and cut off levee for steamboats. But Ecklor House continued to attract river drivers and raftsmen and, in early 1900s, tourists began to arrive in fancy new cars as seen at right.

At this point the company decided to do something drastic. It sent Bacon, now being referred to as "Beef Slough Bacon," with 150 drivers and roustabouts to Eau Claire, some of them armed with "shooting irons," and others with peavies and canthooks, to liberate logs held in Eau Claire booms owned by mill owners on the Mississippi. A tense situation followed, and, though no one was killed, it was later called the "Beef Slough War." Much has been written about this episode. A fairly comprehensive account of it appeared in the Eau Claire *Free Press* on May 7, 1868, under a one column headline which begins with the words THE LOG DIFFICULTY. Several subheadings follow, one of which reads "High handed act in violation of law and equity!" One of the last subheadings reads: "Bacon Arrested and required to give bonds of $20,000." Oral tradition has long held that some arrests were made although the headline suggests only one and the story that follows does not even mention it—a phenomenon not uncommon in modern journalism.

The story, slightly abridged, runs as follows:

"For the last few weeks difficulty has been apprehended by the mill men on the Chippewa from one Bacon, who has become quite prominent from his association as agent of the Beef Slough Company, in the lumbering interests of the Chippewa Valley. A bitter feeling of animosity has been wrought up between the combined lumbering interests of the Chippewa Valley and those interests involved in the Beef Slough project, which are generally calculated by resident property owners to be derogatory to the future development and propagation of the manufacturing resources of the valley.

"Mr. Bacon as the acting man for the Beef Slough Company has taken every opportunity—as evinced by his tenacious spirit—during the past winter to cultivate a feeling of deference and animosity between the interests on the one side and the combined interests of the mill men on the other. During the winter he openly and avowedly asserted that he would not submit to the custom of exchanging logs and, to obtain them he would cut every boom on the river, if necessary. The people who have come to this country and settled among us as lumbermen, have been men, almost exclusively of humble means, who have not yet become able financially to reconstruct assorting booms in order that all may get their logs and have consequently relied upon a system of exchange which has been mutually beneficial to all to stock the various mills on the river.

Frederick Weyerhaeuser's portrait and signature from an engraving of about 1900.

"Mr. Bacon, by his threats, has naturally engendered a bitter spirit between himself and the manufacturing interests. Last Friday, with a crew of 150 men, he proceded to carry out these threats by cutting the boom of Messrs. [M.F.] Hodgins & [John] Robson about six miles above the city, which was variously estimated to contain from seven to twelve million feet of logs. Nine tenths of these belonged to different lumbermen in the valley, and one-tenth to Mr. Bacon. Thus it will be seen that, for the glorification of this man's personality, he deliberately carried out a high-handed act against the sanction of all law and equity, damaging our mill men in general to the extent of several thousand dollars simply for the purpose of carrying out his threat, and obtaining as his reward one million feet of logs, and sweeping from the control of the rightful owners ten times that amount, besides damaging Messrs. Hodgins & Robson to the extent of $20,000.

"We understand that Messrs. Hodgins & Robson offered Mr. Bacon the same terms of exchange given to the other lumbermen and despite these liberal propositions he deliberately turned out this vast amount of property, simply to gratify his bulldog tenacity. A portion of his driving crew were furnished with shooting irons for the purpose of enforcing mob law and the programme as laid out by Mr. Bacon was to go through the log booms belonging to mill men with whom he was at personal enmity.

"A report which has gained credence with many is that Mr. Bacon offered to exchange logs with Messrs. Hodgins & Robson but the offer was refused and his only alternate was to cut their booms and take his property out. Messrs. Hodgins & Robson have one million feet of logs in their booms belonging to Mr. Bacon; the first step to be taken in order to ensure an exchange is for them to turn out an equal number of logs to him; next they must give him control of all logs bearing their marks that pass the booms at this place, and the final credit to be given those men by Mr. Bacon is based upon the scale of logs reaching Beef Slough, thus throwing all the risk of logs sticking on sandbars and running off into sloughs and bayous, upon mill men on the upper river.

"We have yet to hear of any other offer made to our lumbermen by Mr. Bacon. Now we challenge any candid man to point out one feature of fairness in this exchange. It is simply a system of highway robbery sought to be carried out by the aid of armed forces and mob law . . .

"On Saturday morning a body of 200 men were organized under the sanction of the sheriff to meet Mr. Bacon and resist the further cutting of booms. Through the persuasive influence of this 'lobby', armed and equipped with a determination to protect the manufacturing interests of Eau Claire, a compromise was offered in which Mr. Bacon was induced to pass quietly by the mills in this locality and not cut any more booms. The men who so quickly responded to the 'call for the troops' to protect the booms were mostly property holders in East and West Eau Claire, a majority of them belonging to the latter town, who felt it a duty they owed their own. They were determined to establish a precedent that the cutting of booms on the Chippewa River was a dangerous undertaking . . . the local sentiment in the valley is at present wrought up to a pitch that threatened the penalty of death to the men who undertake an outrage of this kind in the future . . . ."

Thus a legend was born. No one was actually arrested, apparently, since the sheriff and his deputies were satisfied to allow the boom busters with their "shooting irons" to pass on but not to come back. The sheriff avoided bloodshed by exercising restraint, but he is not given a name in the *Free Press* story and "Beef Slough Bacon" seems to have left the Slough not long after this incident.

Meanwhile, the Beef Slough Company got the state legislature to grant the company a franchise to maintain the works and booms already constructed on the lake between Nelson's Landing and Alma. Eau Claire mill owners tried to block this move, and managed to get a lumber raft into the slough and float it down to the Mississippi to prove that the slough should be subject to the same rules as a navigable stream. This was a stunt, not a practical matter. In 1872, however, the legislature passed a new law repealing all laws requiring that a passage be left open in sloughs and bayous as long as there was a navigable channel open, and that put an end to the controversy.

By the summer of 1872, after another bad year in the log harvest, Beef Slough Company was facing insolvency, and the stockholders began looking for new capital, or a buyer. At a meeting held at Huff House in Winona on September 5, 1872, it was decided to lease the Beef Slough works to another lumber company organized two years earlier in Chicago called the Mississippi River Logging Company, capitalized at $1,000,000. The new company agreed in its lease to take over the debts and assets of Beef Slough Company, and, by 1879, the entire company had been absorbed by M.R.L. Co.

At the meeting held in 1872, the stockholders went a step further; they elected a thirty-eight year old man as their next president, namely, Frederick Weyerhaeuser who had immigrated from Germany in his youth. It was an auspicious beginning, for Weyerhaeuser would go on from here to found an empire in pine and forest products today known as "the tree growing people."

The new president wasted no time in learning what the Pinery in Chippewa County looked like, and he visited logging camps and watched the river drivers and the teamsters driving crosshaul on the landings. He went everywhere, but no matter where he went, he was as much at home in a camp of lumberjacks as in the front parlor of a millionaire. On one occasion he was traveling in the woods with William Irvine, his manager for Chippewa Lumber & Boom, and the two men stopped over night in one of the company camps. Irvine told his daughter, Ruth (Mrs. Oscar Richter) of this experience and in a personal conversation with the author, she remembered it this way: the two visitors had been assigned to bunks in the sleeping shanty and Weyerhaeuser was given the lower, Irvine the upper. After supper it was not long before everyone was in bed, including the visitors, whereupon a chorus of loud snoring started. Neither of the visitors could get to sleep in this symphony of crosscut saws. Finally, one of the heavy sleepers snorted loudly and fell quiet. Weyerhaeuser tugged at the blanket above for Irvine's attention and said, "Villum, tank Got, von is dead."

One of Weyerhaeuser's timber cruisers in the 1870s and 1880s was Edward Rutledge of Chippewa Falls. He told Weyerhaeuser he had spotted one hundred forties (4,000 acres) of stumpage, but he wondered how the company (M.R.L.Co.) was going to pay for it. Weyerhaeuser allegedly replied, "Ed, I vill find the money; you find the timber." Perhaps this exchange was heard in late January, 1889, when the Phillips *Times* reported that Weyerhaeuser and Rutledge had been visiting the city of Phillips.

With Beef Slough under new management and more capital for improvements, there was no need any longer to store logs in booms above Chippewa Falls at the Paint Creek dam. Beef Slough works between Nelson and Alma could handle as many logs as the sawyers could buck up and teamsters could haul to the river banks. One eyewitness to the sorting works below Nelson said it contained "piles, piers and booms until you can't rest."

There were long walking booms, or catwalks, connecting the various storage pockets or pens. There were two main sorting gaps and, as the logs came down the slough, the catchmarkers checked the end marks and bark marks for ownership and scaled (counted) the logs in order to assess the correct charges against the various lumber companies for "boomage," that is, storage in the Beef Slough booms and for sorting the logs and making them into brails. The catchmarkers made an additional hack of some device at the end of the log to show that the log had been checked. For many logs this was probably the third "hack," since catchmarkers at Paint Creek dam had done the same thing after scaling the logs held at the boom works above the dam, and catchmarkers in Eau Claire were doing the same to levy charges for boomage in local booms. An accurate account was thus kept on all logs in all places no matter whom they belonged to.

The company had already built a frame house at Round Hill for the former boom boss. It was now occupied by "Ab" Gilmore. The buildings stood at the foot of Round Hill overlooking the Chippewa River on the west bank. A small, two-story building was also built for crew and transient log drivers. The main crew of Beef Slough Company, however, worked at the sorting pens and coupling works located below Nelson's Landing where the company office and camps clung to the high bluffs along the east bank of the Slough. The plat of Nelson township (Buffalo County) for 1877 shows two camp buildings on one side of a road, and two opposite, one being the office, all in Section 27. Camp I was located about where the swimming pool is presently situated north of Alma, and Camp 2 was located on the flat where the present consolidated high school stands north of Alma on Highway 35. But there were other buildings in Section 27 to house additional employees, and barns for the cattle and horses.

No one, it seems, ever took a picture of the works below Nelson before Gerhard Gesell, a studio photographer, took some, probably in 1889, shortly before the Beef Slough works were abandoned. There does not seem to be any picture available of the long boom which diverted the logs into Beef Slough below Round Hill.

Meanwhile, at the annual meeting of stockholders of the Mississippi River Logging Company held at Huff House in Winona on June 4, 1880, Weyerhaeuser was re-elected president and Thomas Irvine, secretary. This was the eighth time both had been elected and they would continue to be elected until the company was dissolved in 1909. The meeting on June 4th was probably one of the most important ever held. Word must have reached the members about the great flood raging on the Chippewa River which was smashing booms and knocking over dams and scattering tens of thousands of logs along the Chippewa River.

At the Little Falls dam the embankment on the east end washed out. When the Paint Creek dam, already damaged by the high water, was hit by the extra crest from Little Falls, several gates caved in. In Chippewa Falls the flood removed part of the city dam and damaged the big sawmill. In Eau Claire the new Dells dam held, but the water rose to the top of the deck and thousands of logs splashed over and fell into a foaming mass of angry water. Houses were carried away or undermined, and all the way down the river the logs lashed at the embankments, tearing loose more sod and creating more silt. The effects of the flood were felt all the way to the mouth of the Chippewa River, and Ecklor House at Ella, which had been accessible to steamboats, was suddenly cut off from the river by silting and newly created sandbars.

In modern parlance, the Chippewa Valley was a disaster area. But this disaster had a silver lining, for it helped, finally, to bring the warring chiefs in the logging industry of the valley and on the Mississippi to reconsider their counter productive feuding. In the flood were tens of thousands of logs hopelessly mixed up. It would take weeks to retrieve them and send them on to their rightful owners. It was realized now that the Chippewa River could no longer be trusted, no matter how high the dam or how strong the jam piers. If a flood this bad could happen once, it could happen again. A new order would have to be established to end this chaos.

Into the breach stepped Frederick Weyerhaeuser with a bold plan. He suggested to the feuding mill owners, up and down the river, that they unite in a common approach to the harvest of trees, and that they drive the logs down the rivers as one concern, not as competing companies. Able assistance in this endeavor came from Roujet D. Marshall, a young attorney at Chippewa Falls.

The bargaining went on for several days in Chicago and out of this came the so-called "Chippewa Compromise" signed June 28, 1881. Twelve Mississippi River companies and six mill owners on the Chippewa signed. Knapp, Stout & Company, having an empire to itself, did not join. Under the terms of the agreement mill owners on the Chippewa got 35% of the trees harvested every year, and the mill owners on the Mississippi got 65%. These mill owners made up the membership of the Mississippi River Logging Company whose principal companies included Youmans Brothers and Hodgins of Winona; Laird, Norton of Winona; Winona Lumber Company; W.J. Young and Company and C. Lamb and Sons, both of Clinton, Iowa; D. Joyce of Lyons, Iowa; Dimock, Gould and Company of Moline, Illinois; Weyerhaeuser & Denkmann, and Rock Island Lumber and Manufacturing Company, both of Rock Island; Musser Lumber Company and Hershey Lumber Company, both of Muscatine, Iowa; and Shulenburg & Boeckeler of St. Louis, Missouri.

The six Eau Claire mills who joined "the pool," as it was quickly dubbed, were the Eau Claire Lumber Company, Empire Lumber Company, Valley Lumber Company, Northwestern Lumber Company, Daniel Shaw Lumber Company, and Badger State Lumber Company.

Chippewa Logging Company was directed to handle the affairs of the pool. The timber still standing in Chippewa County and future Rusk County, according to the plat of 1888, was held by a handful of companies, mainly, Chippewa Logging Company, agent for the pool, Chippewa Lumber &

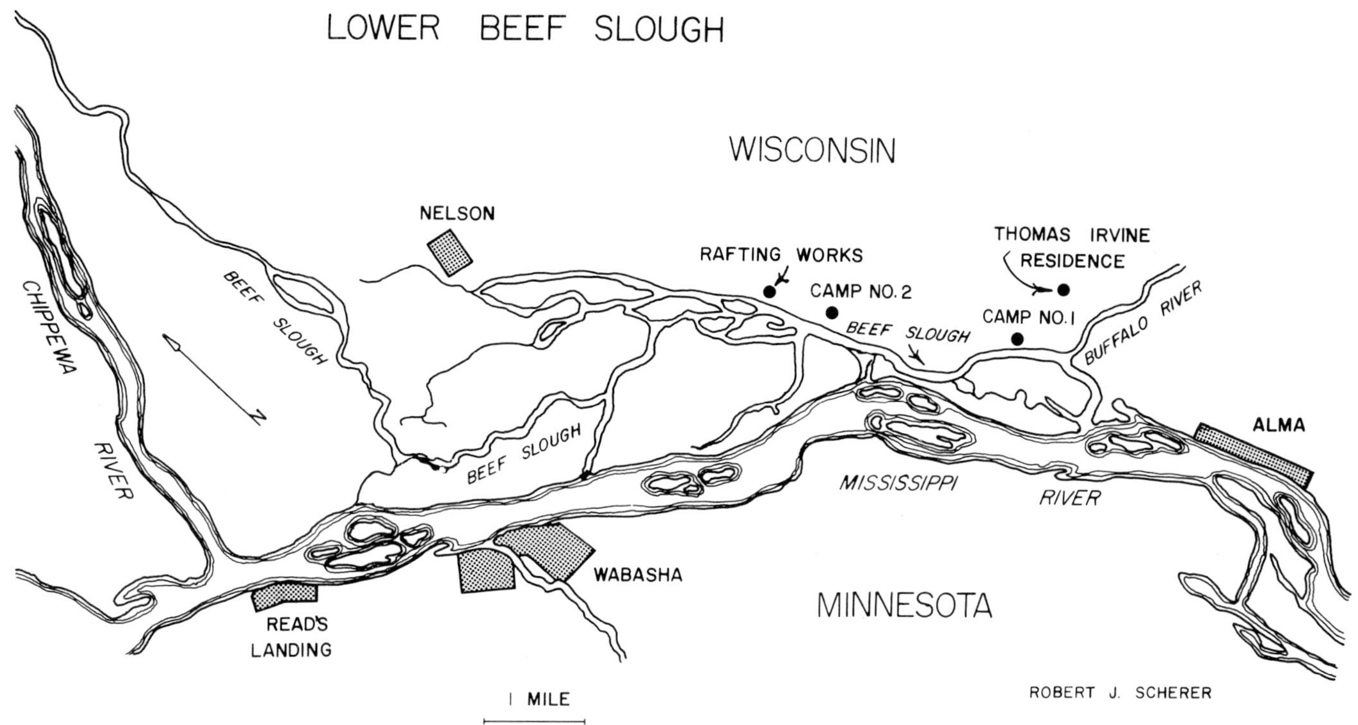

Boom, Laird, Norton & Company of Winona, Daniel Shaw of Eau Claire, and H.W. Sage & Company, an investment firm directed from Sage's home at Syracuse, New York. D.S. Viles, one of the engineers on the Little Falls dam in 1885, also held considerable stumpage in the two counties.

Although there were inequities, the system worked. It was superior to the chaos of pre-flood days when drive foremen, for example, in order to get their logs down the river first, often fought with one another.

The pool's supply of logs came from its own timber lands, from its membership, and from smaller loggers. The pool sent its own crews into the woods when necessary, but depended mostly on contractors. The scaler and his assistants became the pivitol figures in the pool. Only by a minute accounting of every foot of logs cut and measured could the pool succeed.

A committee set prices on the logs in spring according to their quality and location, and each member of the pool received credit for the total amount cut. The most common complaint heard was that some members got better logs than others. The bylaws provided means of arbitration.

Chippewa Lumber & Boom Company was organized in 1879 by businessmen from Chippewa Falls. In 1881, it was purchased by Weyerhaeuser. The big sawmill at the Falls and all the company's stumpage to the north were bought for $1,275,990, which was a bargain. Eau Claire mill owners bought shares in the new company. Weyerhaeuser became president, Orrin H. Ingram of Eau Claire vice-president, and William Irvine, secretary. And the long-ailing sawmill at Chippewa Falls came to life and hummed day and night until the last pine log was sawed for the company on August 21, 1911.

Chippewa Lumber & Boom also rebuilt the damaged Paint Creek dam, raising it higher and, with eleven gates, creating a flowage extending all the way up river to Eagle Rapids. The Eagle Rapids dam washed out in 1877, a second time, and after that no attempt was made to repair it.

Meanwhile, life among the loggers on the Chippewa proceeded apace and with little change. There was talk of using a new type of crosscut saw for sawing down the trees instead of chopping them down, but the change was to come gradually, and a photograph from Minnesota, allegedly taken in 1880, shows two men still chopping a tree, not sawing it down.

The daily journal kept at the Little Falls dam, beginning in 1882, shows that the men were busy from spring to late autumn sluicing logs, watching the dam, and repairing its leaks. All was well. And then in September, 1884, disaster struck again. Heavy rains flooded the Chippewa Valley, and this time the great dam at Little Falls did not wash out around the embankment; the overloaded reservoir smashed it, first one end and then the other. Beef Slough Company, which built the dam, reached a decision to rebuild at once.

A stringer correspondent in Eau Claire for the *Daily Wisconsin* (Milwaukee) wrote that the unprecedented floods "which resulted so ruinously for hundreds of property owners along the Chippewa River" had also resulted in stirring up considerable feeling against the pool companies whose dam, he charged, was a "collosal obstruction to the natural course of the river." (The Pepin County *Courier* reprinted this story September 25th.)

The *Courier* also carried a story on the flood in its issue of September 19th. This report said the river reached its crest on the 11th of September. The daily journal at the dam reveals that the dam went out on the afternoon of September 17th. This need not be a contradiction. It seems that while the flood reached its crest on September 11th, the dam at Little Falls held out for another week, no doubt because of its enormous storage capacity, the equivalent of three square miles.

The *Courier* on September 11th, reported that the sheer boom at Round Hill had torn loose and drifted downriver, but held long enough "to send all the logs that came down safely into the slough." The *Courier* also said that the flood had "swept away bridges, dams, lumber, logs, houses . . ." One of the bridges was the main bridge, built only the year before, between downtown Eau Claire and the West Side. It crashed into the Omaha Railway bridge which also went down.

Again people picked up their shattered lives, even as they had done in 1880, and started over, and farmers who lost cattle in the flood had to look elsewhere for replacements. Like the lumberjacks in the woods who had no insurance, people along the river had no insurance to cover their losses from flood, losses which were largely caused by the destruction of two logging dams owned by private corporations.

Throughout the 1870s and 1880s, officers of the U.S. Corps of Engineers were authorized by the Congress to improve navigation on the Chippewa River, but the funds for these improvements were palliative, and as a result, attempts to control the vast erosion of the banks along the river from the mouth to "Mary Dean" banks (Meredian) were ineffective. Jetties, shore revetments, brush dams and wing dams were built to funnel the river into a smaller channel to carry off the silt and sawdust to the Mississippi, but they were never enough, and resident engineers complained incessently about the lack of funds, and their lack of authority to improve navigation on the river.

A Wisconsin Congressman, W.T. Price, failed to secure an appropriation in 1885 for the Corps of Engineers to make further improvements to the banks of the river, and in a letter to one of his irrate constituents he wrote: "I have not for a day forgotten the matter, nor neglected any opportunity to get justice done to the Chippewa River, but I do not *own* this Congress, and while I am willing to be held responsible for *Effort*, no man can be responsible for *Results* in this mob."

Charles Wanzer, an officer in the U.S. Army Corps of Engineers in charge of conservation on the upper Mississippi and Chippewa Rivers, listed the main dams in the Chippewa Valley in 1878, based on information furnished him by the Beef Slough Company, in this order:

	Height of dam	Flowage
Lower Elk River	10 feet	1 square mile
Upper Elk River	(unknown)	
South Fork of Flambeau	10 feet	4 square miles
East Fork of Chippewa	9 feet	2 square miles
West Fork of Chippewa	9 feet	480 acres
Paint Creek	(unknown)	
Chippewa Falls	(unknown)	
Eau Claire	(unknown)	640 acres
South Fork Eau Claire River	10 feet	320 acres
North Fork Eau Claire River	22 feet	1000 acres
South Fork Eau Claire River	20 feet	800 acres
Little Falls	17 feet	3 square miles

The Dells dam at Eau Claire was probably not completed when this report was made, and the "Eau Claire dam" in the above no doubt refers to the first dam on the Eau Claire River near the sawmill of the Eau Claire Lumber Company north of the Dewey Street bridge. It remains a mystery, though, why there are no figures for the dam at Paint Creek, but no mystery why the Eagle Rapids dam is not mentioned: it had been abandoned.

With all the silt and sawdust coming down the river, it was probably a fair guess that the entrance to Beef Slough, where there was already a sandbar,

Round Hill

Dear Sir

If you have plenty room in the Slough, Please write or telegraph us to Chippewa Falls. We can & will make a flood — providing you and Mr Irvine say so.

Yours very truly

F. Weyerhaeuser

Above, Albert ("Ab") Gilmore, boom boss at Round Hill.

Left: probably one of handful of letters ever written from Round Hill, this one by Weyerhaeuser addressed to "Ab" Gilmore: "If you have *plenty room* in the slough, please write or telegraph us to Chippewa Falls. We can & will make a flood—providing you and Mr. [Thomas] Irvine say so."

would grow into a bigger sandbar. The Beef Slough Company had asked the Corps of Engineers to dredge it and offered to pay for the dredging, but the offer was rejected.

A report by an officer of the Corps of Engineers for 1889 gives the lineal feet of shoreline at various points on the river which had been improved in one way or another, and since most of the place names are no longer used, it seems well to repeat them here, beginning at the mouth and working upstream, namely, Little MO Chute, Flower Pot Island, Plum Island Flats, Plum Island, Dead Lake Cut Off, Three Mile Prairie, Five Mile Bluff, Twin Islands, Durand, Waubeek, Yellow Banks, Dark Slough, Wacouta Island, Jack Staff Island, Mary Dean, Rumsey's and on to Eau Claire.

All told, about $200,000 was expended by the Corps of Engineers from the early 1870s to 1893. At the end of this time, a resident officer on the Mississippi admitted that the only place where anything permanent had been accomplished was at the mouth of the Chippewa.

Robert F. Fries, writing in the *Mississippi Valley Historical Review* (Vol. XXXV), says that government funds were wasted on the river "as long as the logging companies continued to obstruct traffic."

The Beef Slough works, despite all the money spent on improvements, were still unable to handle the ever mounting flow of logs from the north. Faced with this, and the unending struggle against silting at the mouth of the slough, Weyerhaeuser decided to take his logs to the Minnesota side of the Mississippi River and abandon Beef Slough entirely. There was a slough at West Newton, about half way between Wabasha and Winona, Minnesota, which had been used on a small scale as a storage and rafting facility. With new improvements and more booms, it could handle any number of logs.

The Pepin County *Courier* carried the obituary on Beef Slough on October 17, 1889. Without a headline but buried among local items written by a stringer in Alma, this appeared: "The Beef Slough Co. has built extensive works at West Newton, Minn. this fall including offices etc. preparatory to a transfer of the business to that place. Beef Slough is no more."

The slough was still there but not the company which, in fact, was not officially dissolved until February 16, 1905. But it had, to all purposes, been replaced before it died by a new organization incorporated on April 6, 1889 as Minnesota Boom Company. The company was organized to erect booms, sorting and coupling works, piling and piers, and, apparently, to install a sheer boom below the mouth of the Chippewa to shunt the incoming logs as far into the middle of the Mississippi as possible toward the Minnesota side. Whether this sheer boom was ever erected appears doubtful, but the intentions of the boom company to build piers and booms seemed to be in violation of existing navigation laws governing the Mississippi River. Fearing opposition the company went ahead with its plans without delay and leaked the rumor to Major C. Mackenzie, Corps of Engineers, that if the sawmills did not get their logs, the mills would be shut down and throw 12,000 men out of work.

The next fifteen years, that is, roughly 1890 to 1905 when the last log for the M.R.L.Co. came down the Chippewa, were highlighted by law suits and legislative resolutions directed against the Weyerhaeuser-led companies for alleged violations of federal navigation laws and defiance of state laws. For example, the federal government had built a dam some years earlier across the head of West Newton chute which was now in the way of the company's logs entering the chute. Kenneth McKenzie, boom boss at West Newton, had one of his demolition experts blow a section of the dam out with dynamite. This could not be considered a misdemeanor; it was a felony. The case was heard before a court in La Crosse. The foreman testified that he acted on his own initiative. When this information later reached the secretary of the company, Thomas Irvine, the latter allegedly told McKenzie he should not have done it but now that he had, he was not going to blame him for it. But, after hearing several witnesses, the grand jury dismissed the case for lack of jurisdiction. The offense had been committed on the Minnesota side. The court, however, warned the company not to blow out any more dams but made no move to punish the foreman. The company offered to pay for the damages, but there is no evidence that it did or that the break in the dam was ever repaired.

The mounting arrogance of the Minnesota Boom Company led an assemblyman from Chippewa Falls to introduce a resolution in the 1891 session of the state legislature relating to alleged violations of navigation laws on the Mississippi River. Without naming any names, he asked the Secretary of War, who commanded the U.S. Corps of Engineers, to remove from the upper Mississippi "any and all piers, booms or other works heretofore placed there without authority of law, by any private corporation, or persons, which in any manner interfere with the free navigation of said river . . ."

The resolution was duly passed in Madison and sent on to Washington where Congressman Allan R. Bushnell introduced a resolution of a similar nature in the House. From there it was passed on to the Secretary of War who passed it on to the Corps of Engineers who sent it back to the resident engineer, Major Mackenzie, at Rock Island, Illinois, who wrote a report on it and sent it back to his superior who sent it on to the Secretary of War who sent it on to the Speaker of the House where the buck finally stopped and it became a matter of public record in House Executive Document No. 183, 52 Congress, 1st session, titled "Government Dams on the Mississippi River."

Meanwhile, the secretary of the Mississippi River Logging Company had sent a letter in June 1890 to the Secretary of War requesting approval to install sheer booms and crib piers below the mouth of the Chippewa, quoting legislation enacted by the Congress on June 16, 1880 as his authority, but the Secretary of War rejected the request, and boom piers were liable to removal under legislation enacted September 19, 1890 "if shown to be unwarranted obstruction to navigation."

A series of letters were exchanged between McKenzie and Thomas Irvine, the company secretary. In fact there are nineteen pages of correspondence in Document No. 183, none of which stopped any logs from running loose on the Mississippi River. The letters from McKenzie suggest that he was only lukewarm to any suggestion to halt the flow of logs, which, of course, is what Irvine wanted.

There was a small hitch, however. Minnesota Governor Knute Nelson on April 7, 1893, appointed John M. Mullen surveyor-general of the 4th District on the Mississippi River, a district which began at a point below the outlet of Lake Pepin south to the south line of Wabasha County, an area which included West Newton slough. Section 2400 of Minnesota statutes for 1894 made it plain that it was his duty "to survey [scale] all logs and lumber running out of any boom" in his district. The efforts of Mullen to carry out his duty in the face of the evasive and often violent tactics adopted by Minnesota Boom Company would have tested the patience of Job.

One of the foremen at West Newton, probably George Scott, told Mullen he could scale the logs on the outside of the boom works which did not belong to the company pool. His own company's logs from Wisconsin, he made clear, had already been scaled and there was no need to scale them a second time and pay five cents per thousand (about five cents for two 16-foot logs). Mullen was further advised that if he persisted in scaling the company logs, he would have new problems to deal with unrelated to scaling.

On Sunday, April 29, 1893, a barge was loaded with lumber at Wabasha for West Newton slough where Mullen intended to build an office. At the slough his men were met by men from the local boom who cut the lines and sent the barge adrift. The Wabasha *Herald*, on May 4, 1893, asked "what is the object of all this unless to prevent General Mullen from doing his official duty as prescribed by statute? Is it, or is it not, evidence that the company, with their millions back of them, are trying to defeat the law by making those whose duty it is to enforce it powerless . . . ?"

The following week, the *Herald* reported that the clash between Mississippi River Logging Company and Mullen "is about to reach a focus and in a short time all will know whether the logging company or the Minnesota statute is the stronger."

Fred W. Kohlmeyer, author of *Timber Roots, The Laird, Norton Story*, pays no attention to the charges made in the Wabasha *Herald*. Instead, he says a committee of lumbermen went to Mullen and found him to be "arrogant, in constant state of temper, and with no conversational restraint whatsoever."

Mullen, who was not a general of the army but a surveyer-general, now took his case to court and got an injunction against further interference, but the Wabasha *Herald* charged that the company only pretended to obey the injunction and ran its logs out before they could be scaled and forced Mullen to chase after them downriver. The *Herald*, concluding its editorial of May 11th, had this to say:

> It is a fight of a great monopoly against the powers of the state—the managers of the Mississippi River Logging Company and the Minnesota Boom Company have thus far played successfully a game of bluff against the land and log owners of the Chippewa Valley; they have transformed the Mississippi from Read's Landing to Winona from a noble interstate highway to 'Weyerhaeuser's Ditch,' but now, for the first time, they have met a foeman worthy of their steel.

Not quite. The company was not giving up that easy. An armed truce followed while Mullen and his assistants scaled what logs they could get to without being tossed into the river by the boom crew. At the end of season, he sent Minnesota Boom Company a bill for his services which the company ignored. He then made an attachment on a number of log brails owned by several companies. This action was at once replevined by the owners, that is, legal action was initiated to recover the logs which they insisted had been wrongfully attached.

The case was heard in United States District Court in St. Paul, and Mullen's attorney won a verdict in his favor. The plaintiffs, Lindsay & Phelps, promptly filed notice of appeal and the suit of Lindsay & Phelps Plff. in err v. John H. Mullen and the State of Minnesota went to the United States Supreme Court where the case was argued in the October term of 1899. With four justices dissenting, the decision of the lower court was upheld and six companies were assessed a collective fine of $27,669.99, namely, Lindsay & Phelps, Mississippi Towing Company, John Paul Lumber Company, Davidson Lumber Company, both of La Crosse, Wisconsin, all non-members of the Mississippi River Logging Company, and Hershey Lumber Company and C. (for Chancy) Lamb & Sons, both affiliated with M.R.L.Co.

Lindsay & Phelps, a towing company from Davenport, Iowa, contended that they could not be held responsible for scaling logs which did not belong to them. They were in towing, not logging. On the other hand, it seems remarkable that the company was the plaintiff in a case involving a scaler whose duty it was to scale all logs leaving West Newton slough. Nevertheless, here was a suit which was argued before a district court in St. Paul and before the Supreme Court of the United States and yet is not mentioned in the standard biography of Frederick Weyerhaeuser called *Timber and Men: The Weyerhaeuser Story* published in 1961. Instead, the authors make the following comments:

> The rash of damage suits and the dispute as to the scaling of logs at West Newton became disastrous publicity for Weyerhaeuser and his associates. Millions of Americans heard of them, never ascertained the truth, and assumed that they indicated grave abuses instead of the relatively trivial litigations, often lacking substance, which they actually were.

Meanwhile, the state of Minnesota, taking cognizance of the drain on the purse of its surveyor-general, appropriated $15,800 for his relief. But it took nearly seven years for the appeal to be heard in the Supreme Court, and since no one knew how the court would adjudicate, Mullen, casting about for friends, suddenly found himself standing alone on the West Newton boom. The attorney-general in St. Paul looked the other way and saw no evil. Mullen was forced to compromise with Minnesota Boom Company or lose his job. An accommodation was reached and a year after the appeal was entered, that is, 1894, a boom company employee was made a deputy to Mullen. Throughout the last years of the decade, the scalers at West Newton had to have the tacit approval of Minnesota Boom Company and this was confirmed when it was learned that one scaler, looking for a job, followed Weyerhaeuser all the way to Chippewa Falls for his endorsement.

In a series of articles written for the Burlington (Iowa) *Post* in 1930 by Fred A. Bill, who spent at least four decades on the Mississippi River in mills and steamboating, the Minnesota Boom Company is charged with "deliberately and knowingly" violating state law, and the boom company, "a creature of the state, did everything in its power to prevent the enforcement of the law . . ."

For many years a ferry operated between Wabasha and Nelson's Landing on the Wisconsin side, but with the increasing flow of logs floating across the Mississippi to West Newton slough, the ferry was constantly in danger of ramming a log. Citizens sent a picture of the river, cluttered with logs, to an agency of the government, and months later, Minnesota Boom agreed to build a pontoon bridge. But floating logs made it unsafe and boat owners, ferry riders and farmers "lingered at the shore and damned Weyerhaeuser and his logs."

The peak year at West Newton was reached in 1892 when 632,150,000 feet of logs were rafted into brails and sent downriver. The last drive of Mississippi River Logging Company on the Chippewa included more than fifty million feet of pine which was run in the spring of 1905. Fred W. Whitney, a cook on the wanigan, told Fred Bill that he arrived at West Newton with the last logs and was paid off with the rest of the crew on June 4th. Will Grippen, tender on the draw bridge at Durand, thought that the *Good Luck*, a combination steamboat and pile driver, towed the wanigan back to Eau Claire the last days of May, and about June 7th, the last stray logs were sacked into the river. Jacob Bert, one of the foremen at West Newton in 1905, thought the last logs were coupled into brails on June 19th and pushed downriver to the sawmill of Weyerhaeuser & Denckmann at Rock Island.

On August 16, 1905, the piling and boom sticks at West Newton were sent downstream in a sack boom to the Laird, Norton sawmill at Winona, and the company office, cook shanty and sleeping shanty were sold on the spot for scrap lumber. And that ended the saga of "Weyerhaeuser's Ditch."

Railroad Logging

On a television documentary one evening, Grandfather was telling Grandaughter what he did when he was a lumberjack. In the course of the conversation he mentioned a logging train. Grandaughter wondered what a train was doing in the woods. She thought trains ran only through towns and carried people, not logs. But Grandfather was right.

Logging trains were operated in the woods both on trunk lines and on spurs built off the trunks, but there were no two alike. Some were narrow gauge spurs and some were standard gauge.

Spur running into camp of West Lumber Company, Flambeau township, Price County.

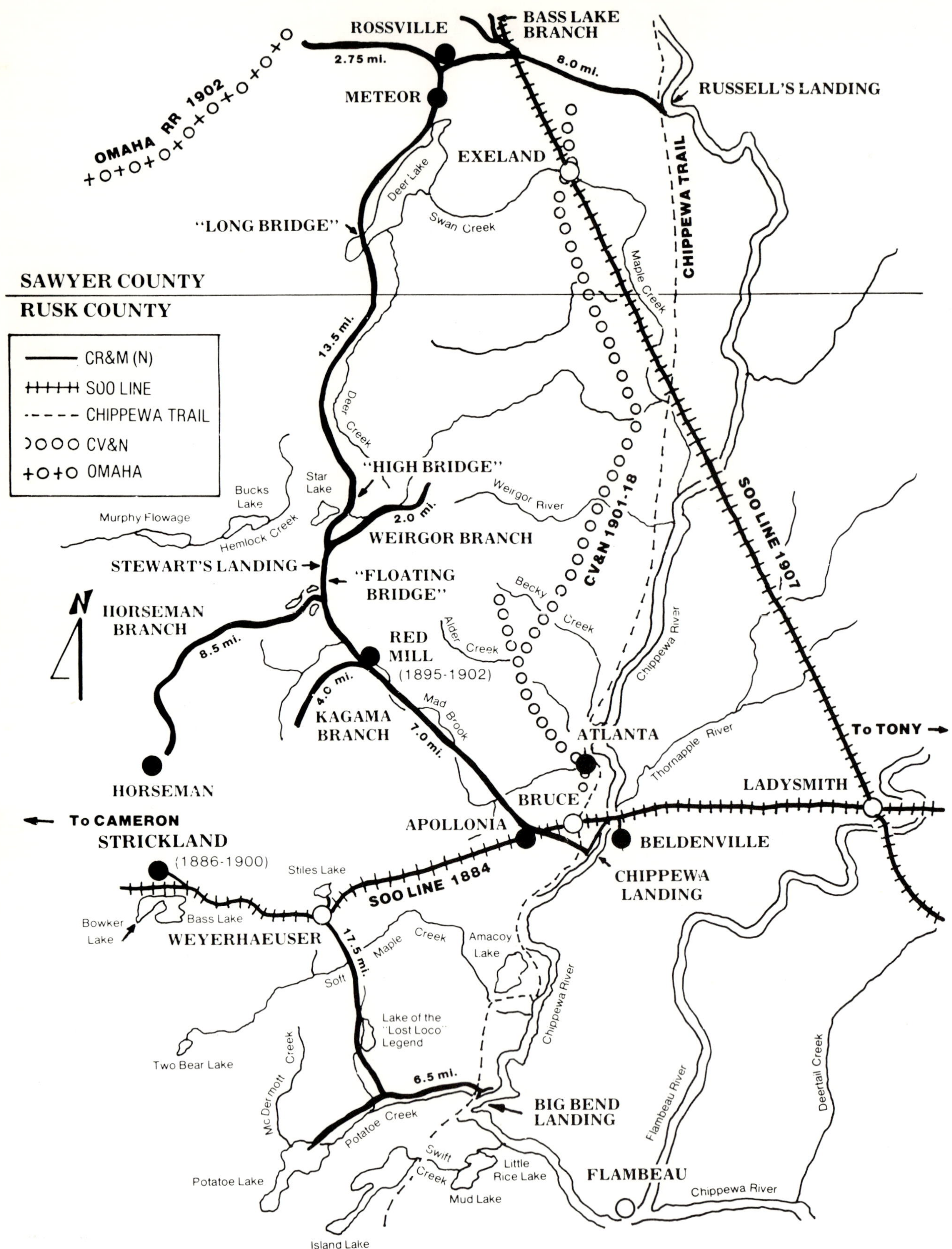

Map of Chippewa River & Menomonie Railway from *Rails Into the Pines* (1980), reprinted by special permission of the author, R.C. Brown (Eau Claire, Wisconsin). The spelling "Potatoe" used on map follows original spelling in early documents related to area.

Everything was done according to circumstances and financing available. Take, for example, the Chippewa River & Menomonie Railway. This was laid in the middle of the wilderness with no connection to a trunk line. It was built to bring logs from camps around Potato Lake over to Big Bend on the Chippewa River. Here the logs were decked into enormous rollways and when spring came, they were rolled into the river and floated down to Chippewa Falls, or all the way to the Mississippi River.

Railroad logging never played a big role on the Chippewa. There was little need for it; the Chippewa River and its tributaries were the equivalent of a complex railway system. To the west of the Chippewa Valley, however, railroad logging became important around Shell Lake where there were no rivers which could bring logs to a sawmill. Instead, spurs were built into the white pine belts by the Crescent Springs Railroad to haul the logs to a sawmill built by Weyerhaeuser and associates on Shell Lake. Trackage was laid down to a grove of pines and after the pine was cut, the tracks were pulled up and a spur laid down into another grove, using the same rails and often the same ties.

The actual trackage of logging railroads in the Chippewa Valley up to 1885 did not exceed eighty-eight kilometers (55 miles).

Loading operations on cars, usually Russel cars, were done in the 1880s and 1890s with single chain. But around the turn of the 20th Century steam loaders were introduced such as the McGiffert which was manufactured in Duluth by the Clyde Iron Works. In some picture captions the loader is referred to as a "hoisting machine," since it hoisted the logs into the air and onto the cars or sleighs. The success of the McGiffert loader, and earlier models, led to the adaptation of the portable woods jammer. This was a gin pole at least ten meters long hovering over a sleigh or railway flat car equipped to load logs with a team of horses on "cross-haul." There were several models of the woods jammer developed in the first decade of the 20th Century, but all served the same purpose: to load logs with horses. (See accompanying photos.) The woods jammers could also be moved on a pair of skids around the woods by the horses, and, finally, they could be manufactured locally by the camp blacksmith and carpenter.

There were two logging railroads laid in the Chippewa Valley almost at the same time, namely, the Chippewa River & Menomonie, which was built in the winter of 1882-83, from Potato Lake to Big Bend on the Chippewa River, and the other, which was laid east of Phillips, probably in 1883-84, called the Lake Shore & Eastern Railroad of the Phillips Lumber Company.

The Chippewa River & Menomonie was organized at a meeting of stockholders of the newly-incorporated Chippewa Logging Company, agent of the "pool" organized by Weyerhaeuser. The logging company, according to R.C. Brown, author of *Rails Into the Pines* (1980), had purchased 111,000 acres of stumpage from the trustees of Cornell University, an educational institution in New York state which had been granted Wisconsin pine lands. The problem was: how to get the logs to the Chippewa River? When the purchase was made it was thought that the logs could be driven down Potato and Soft Maple creeks and Mad Brook (later Devil's Creek), which empties into the Chippewa above Big Bend. But Potato Creek was hard to drive despite help from two dams and a third dam on a branch called McDermott Creek.

Work on the Chippewa River & Menomonie was begun in the late autumn of 1882 and continued through the winter, the track layers progressing eastward from Potato Lake to Big Bend, a distance of eight miles. Two locomotives were purchased for the line, the first one, a saddle-tank dinkey made by the H.K. Porter Locomotive Works in Pittsburg, was hauled on the Omaha line up to Bloomer and from Bloomer sleigh-hauled on the Chippewa Trail to Potato Lake.

In the winter of 1883-84, a branch was built to a junction on the Soo Line and not long after a community developed near the junction called "Weyerhaeuser." The branch, when completed, measured fifteen and a half miles. This junction with the Soo Line also made it possible to bring the second locomotive into the woods without having to sleigh-haul it from

Photograph taken from postcard dated Winter, Wisconsin, March 14, 1914, shows geared engine used by Hammond & Chandler Lumber Company in Sawyer County. Eleazer S. ("Lee") Hammond was from New Brunswick, Canada, and worked in Maine woods before moving to Eau Claire in 1867. The Hammond-Chandler Lumber Company was organized in Rice Lake, and later became Hammond-Olson Lumber Company. Both Ray C. Chandler and Abner Olson married daughters of Hammond. The Phillips *Times* on February 16, 1889, reported that "Lee" Hammond was expected to bank 12,000,000 feet of timber, presumably in Price County, that winter.

Bloomer in the manner of No. 1.

Construction costs of the C.R. & M. railway were $32,400 for the main line between Potato Lake and Big Bend; $32,500 for the Weyerhaeuser branch, and $2,600 for spurs west of Weyerhaeuser. The value of this short stretch of steel to the lumber company may be judged from the fact that in one year alone, 1885, a total of thirty five million feet of pine was landed at Big Bend.

After the pine was harvested south of Weyerhaeuser, according to Brown, the company moved north on the Chippewa River to establish "Chippewa Landing" which lay a short distance below Bruce. Here trackage was laid down in 1887-88 along the river overlooking the high banks on the west side. Equipment used on the original line at Potato Lake was moved over the Weyerhaeuser branch to the Soo line and east over the Soo to Verona Junction, today Apollonia, connecting with the track along the river front. Appolonia became, in a sense, a "division point" and eventually had an engine house, shop and depot.

Chippewa Landing now became the center of logging in northern Chippewa County (today part of Rusk County). A camp was built for the crew which unloaded the cars here and in spring rolled the logs into the river. The Daniel Shaw Lumber Company of Eau Claire was also logging around the east slope of the Blue Hills and hauled its logs over to Chippewa Landing on C.R. & M. trackage.

On September 1, 1902, with no more pine in sight nearby, the Weyerhaeuser group closed out its operations around Chippewa Landing and sold what could be salvaged of the C.R. & M. to a new company organized the same day called the Chippewa River and Northern Railway, W.H. Phipps, president, S.C. Phipps, vice president, W.J. Pierpont, treasurer, and C.P. Coon, superintendent. One locomotive and probably fifty log cars were included in the deal. The new railway was used mostly to haul hardwood which was being sawed at a mill of the Beldenville Lumber and Veneer Company, east of Bruce. A new 4-4-0 locomotive was purchased, the fifth one to be used since the C.R. & M. was built. For a time, Menasha Woodenware was also bringing in hardwood to the company mill in Ladysmith over Chippewa River & Northern tracks.

The railroad era in the Chippewa Valley actually began with the arrival in Eau Claire on August 11, 1870, of the first passenger train on the West Wisconsin Railroad from Milwaukee. A big celebration with a street parade and free lunch marked the occasion. The railroad was extended to

Big Bend landing on Chippewa River, about 1885. Logs were brought here from Potato Lake and camps to west around Blue Hills, and decked into rollways pending spring when logs were rolled into Chippewa River and floated down to mills. River at right is covered with ice. Engine No. 2 for Chippewa River & Menomonie Railway has just arrived with log cars. At upper right appears to be camp for crew stationed on landing.

Small caption at bottom of picture reads: "C.&M. Co. Logging Train at the Landing." Actually it is Engine No. 2 (2-6-0), about 1885, arriving at Big Bend with twenty-seven Russel cars and caboose.

St. Paul, and in June 1875, the Chippewa Falls & Western Railway was laid from Eau Claire to Chippewa Falls. This line was later taken over by the Wisconsin Central.

The Chippewa Valley & Superior Railway, incorporated on June 15, 1881, built forty-six and forty-seven tenths miles of track from the Mississippi River to Eau Claire, and sold out the following year to the Chicago, Milwaukee & St. Paul railway. The Chippewa Valley line also built a twenty-mile branch connecting Red Cedar Junction with Cedar Falls, north of Menomonie, and this was also taken over by the Milwaukee Road in 1882.

The Wisconsin Central railway, in 1872, laid trackage from Colby and Abbotsford as far north as mile point 101, also known as Worcester Station which was located about eight miles south of present Phillips. But here, for lack of funds, the company discontinued track laying and it was not until 1876 that work was resumed and the line extended to Park Falls and Ashland. The railway company entered a town plat on September 23, 1876 at what was then Elk Lake, but which was now changed to "Phillips" to honor Elijah B. Phillips, one of the construction engineers who built the line up from Colby and Abbotsford.

The Omaha railroad, in 1902, connected Holcombe with Chippewa Falls via Jim Falls and Cornell on the east bank of the Chippewa, and at the same time, built a branch line east from Holcombe via Arnold and Donald to Hannibal to bring out the timber and lumber from mills and camps along the line. A spur also went south from Arnold to Ruby where the Ruby Lumber Company built a sawmill that year in anticipation of a railroad. The spur from Arnold was called the Wisconsin, Ruby & Southern Railway.

The Fountain-Campbell Lumber Company built a sawmill at Donald. With the lumber from the two mills at Donald and Ruby and the logs picked up along the line, the Holcombe-Hannibal branch proved a boon to mill men and logging contractors. The crew which came up from Chippewa Falls brought the empty cars over to Hannibal and returned with as many as thirty cars a day of logs and lumber. But this bonanza lasted only a few years before the pine disappeared and finally the hardwoods.

The Minneapolis, St. Paul, Sault Ste. Marie railway was laid from Turtle Lake to Bruce in 1884, and trackage went later to Warner, a name which replaced Flambeau Falls which in turn was replaced by Ladysmith.

With all the new railroad connections bisecting the Pinery, it was now possible to get fresh vegetables and fresh meat into the logging camps. And this changed the quality of the food served, and cooks began to wear white aprons instead of blue denim around their midriff.

A second important logging railroad in the Chippewa Valley was built in Price County in the mid-1880s called the Lake Shore & Eastern, a line which ran along the south shore of Lake Duroy in Phillips and east, eventually as far as the Willow River. The Phillips *Times* reported that the railroad had two locomotives, fifty cars and thirty miles of track.

The line was built by the Phillips Lumber Company but it was never incorporated as a separate company. Phillips Lumber Company was incorporated on July 9, 1883 for the purpose of "buying and selling of pine lands and the manufacture of and dealing in lumber." One thousand shares at $10 a share were issued by the incorporators who were John R. Davis, John Q. Griffith, Carver N. Griffith, and Benjamin W. Davis.

The Davis family was from Neenah, and the Griffiths from Fond du Lac, Wisconsin. It is generally accepted as a fact that the partnership between the two families was agreed upon some time in 1882 when the principals came to Phillips to look for timber and an opportunity to saw lumber. However, not long after the incorporation was legalized, the Griffith family withdrew, leaving John R. and Ben. W. Davis to their own resources, and out of this came a company and a railroad which were to dominate the life and times of Price County for nearly three decades.

The company may have been the first in Wisconsin to install a "hot pond" below the sawmill. Most mills closed down after the ice froze on the local pond or river and did not resume

This bridge, known as "High Bridge," lay in Rusk County and was 1470 feet long and more than eighty feet high. Rented C.B.&Q. engine 4-4-0 pulls log train, about 1903.

Typical landing with log rollways decked in winter months by horse jammer seen in background. Flat car on track has stakes. Photo probably taken near Altoona.

The Nash Lumber Company of Wood County operated a sawmill at Shanagolden, west of Glidden on east fork of Chippewa, and train seen here is crossing creek which probably empties into east fork.

First engine to enter Eau Claire on West Wisconsin Railway, August 11, 1870, shown later when being used as switch engine for Omaha line.

operations until spring. But by creating a special enclosure in the pond below the bull-slide, and warming the water, frozen logs could be rolled into the water all winter, and the ice, debris and sand dissolved in the water. The Phillips *Times* on November 24, 1888, explained how it worked. "They [Phillips Lumber Company] will keep the lake open by charging the exhaust pipe so that it will exhaust into a condenser near the foot of the [bull] slide . . ." The news item ends with the note that the "John R", a pet name given by the crew to one of the two locomotives owned by the company—after John R. Davis—would "endeavor to keep the mill supplied with logs."

On July 8, 1889, the Phillips Lumber Company was dissolved and all debts and assets were absorbed by its successor, the John R. Davis Lumber Company. The announcement appeared in the Phillips *Times* but no reason is given for the change in corporate status.

By early 1894 the John R. Davis Lumber Company was being referred to in the *Northwestern Lumberman* (February 10th) as "among the greatest of great lumber plants of the Wisconsin pine country."

By this time, a big planing mill and a dry kiln were in operation, but on June 23, 1894, both burned, a prelude to an even greater catastrophe on July 24th when a fire spread from a burning swamp nearby and, fanned by a strong wind, destroyed the entire city, mills, homes, stores and Court House, with a loss of fifteen lives. Box cars and coaches on the Wisconsin Central were pressed into service to evacuate women and children to Prentice, but there was little the fire fighters who remained behind could do to stop the holocaust.

A marker erected by the State Historical Society of Wisconsin on Highway 13 overlooking Lake Duroy calls attention to the fire and after the smoke died away "these same valiant people built a new Phillips on the ashes of the ruins."

The lumber company mills were rebuilt and the most advanced machinery installed. The company also built another railroad, this one south of Fifield east along Sailor Creek. Another spur was built off the main line of the railroad which extended to Coolidge, today a ghost town, where the Boyington & Atwell Company of Stevens Point was sawing hemlock in the late 1880s and early 1890s.

The accompanying map shows a branch line off the Wisconsin Central running northwest from Luger Junction to Lugerville, with short spurs out of Lugerville into the timber, actually Camps 6 and 7 in Flambeau township of Price County. This sawmill community was founded in 1904 by Louis & William Luger who came from Minnesota to build a mill on the east bank of the south fork of the Flambeau River. They picked a spot well suited for sawmill operations directly below Little Bull Rapids, although these rapids are not as well known today as Little Carry and Rocky Carry Rapids located a short distance upstream.

In 1909 the big sawmill at Phillips burned, and the Davis company took over the sawmill of the Luger brothers at Lugerville. The mill was remodeled for greater efficiency, and green lumber was shipped by train to Phillips to be planed and dried. But production declined and in 1912, Davis conveyed the mill at Lugerville and all his mill properties, planing mill, dry kilns, machine shops and yards at Phillips, to Davis Kneeland and Percy McClurg, two out of state lumbermen who moved to Phillips.

A new sawmill at Phillips was begun in the summer of 1910 but construction was delayed by Davis. It was finally completed in February 1913, presumably under new management, and, in order to keep the Phillips mill supplied with logs, new camps were opened as far east as the Willow River in Oneida County (northwest of Rhinelander).

In 1914, George A. West and partners of Milwaukee acquired the Kneeland-McClurg mill at Lugerville and began operations under the firm name of West Lumber Company. The company sawed until about 1933, and Volume I of *Corporations* (p. 191) for Price County shows that the company was dissolved December 24, 1935. James T. Drought was president, and George A. West, secretary, at the time of dissolution. George Heckman was also a member of the firm.

After 1914, Lowrie Lowe was mill superintendent and Herman J. Johnson was manager of the West Lumber Company. In the

1920s Roy Bodenburg was head sawyer on the night shift.

There were two engines, 101 and 102, which ran from the mill to the several camps in Flambeau township to pick up cars. When the company was dissolved, 101 went to a buyer in Mexico, and 102 was junked.

On May 19, 1922, the sawmill at Phillips burned for the third time. It was rebuilt the same summer and the saws began to hum again on November 11, 1922, but the end of the pine harvest was in sight. The Phillips *Times* on June 18, 1926 wrote the obituary, and eulogy:

> After forty-four years of continuous logging operations by the Phillips Lumber Company, the John R. Davis Lumber Company and the Kneeland-McClurg Lumber Company, successive operators of the sawmill plant in this city, the last of the company's great holdings of timber located east of this city will be hauled to the mill and their logging camps will be closed. By the first of August the Kneeland-McClurg Lumber Company expects to completely finish the taking up of the steel rails of their logging railway over which has been moved to their mill during those forty-four years nearly a billion and a half feet of logs, at a conservative estimate. The annual cut of logs has been from twenty-five to sixty million feet.
>
> Established in 1882 as the Phillips Lumber Company, with John R. and B.W. Davis, the controlling managers. A few years later the name was changed to the John R. Davis Lumber Company and in 1912, the name was again changed to the Kneeland-McClurg Lumber Company. John R. Davis the real founder of the industrial enterprise still retains an interest in the business.
>
> This fine lumber manufacturing plant which we understand to be the largest electric driven sawmill in the world, has always been recognized as the industrial support of Phillips and has ever been most liberal in all matters of public improvements that has made Phillips a modern city and furthered its progress and commercial development. The Company has never sought to dominate in municipal affairs, but rather to harmoniously assist in the upbuilding of the city.
>
> The exhausting of the timber supply east of the city is a loss that we will feel, but it does not mean that the sawmill plant will be closed down now, or for years to come. The company's timber holdings in other points, Morse and elsewhere, assures the operations of the plant as heretofore. Phillips, without unforseen conditions arising, will yet remain a great lumber center, and a prosperous city.

The managerial capability of the Kneeland-McClurg combination was severely tested on two occasions. David Kneeland died suddenly of a cerebral hemorrhage at Phillips on December 8, 1915. Percy McClurg kept the company on a profitable basis and by 1923 had become a national figure in the lumber trade. In August he left on a trip to Czechoslovakia with a group of lumbermen from around the country to study the possibility of American investment in Czech forest products. He returned to Phillips in November and a few days later, on December 4th, he was on an ice skating party on Elk Lake, broke through the ice and drowned. Children of the founders and relatives managed to keep the company going for a time, but, as noted in the Phillips *Times*, the end was in sight for no amount of managerial skill could overcome the fact that there were no more trees to cut except pulpwood.

In the late 1890s and early 1900s there are references in photographs to several logging railways in the Chippewa Valley as far north as the Thornapple in Sawyer County. For example, the Omaha line came from Spooner to Winter in the summer of 1904, and, around 1905, there was a siding east of Winter which connected with the "Stinson Spur," which ran south for about fifteen kilometers (ten miles) to the Thornapple River, crossed the Thornapple and continued east and a short distance to Camp 5. The spur was named for Charlie Stinson of Augusta, woods boss for Kaiser Lumber Company of Eau Claire. Camp 4 for Kaiser stood on the right (west) bank of the Thornapple where the bridge was built. The clearing where Camp 4 stood is still an opening with no trees.

Most of the stumpage taken out by Kaiser was hemlock, birch and basswood. The pine had long ago been cut.

Other logging railroads in the Chippewa Valley would include the Stanley, Merrill and Phillips line

Chicago, St. Paul, Minneapolis & Omaha train crossing Chippewa River at Eau Claire, probably September 1, 1880. Caption on car reads "This train loaded with lumber for Omaha, Nebraska from the Eau Claire Lumber Company, Eau Claire, Wis." This was special train of twenty-four "lumber line" cars each thirty-five feet long which carried lumber as well as lath and shingles. In background is mill of Eau Claire Pulp & Paper Company, one of first paper mills in state.

north of Stanley for the Northwestern Lumber Company. Spurs off the Tuscobia branch of the Omaha line for the Edward Hines Lumber Company at Rice Lake, Loretta and Park Falls picked up log cars, and the "Roddis Line" (for Roddis Lumber & Veneer Company of Marshfield) was hauling log cars from the headwaters of the Flambeau River to the company sawmill at Park Falls.

Fountain-Campbell Lumber Company of Donald, later of Ladysmith, laid rather extensive trackage into Rusk and Sawyer counties northeast of Crane. (This company was later absorbed by the Yawkey-Bissel Lumber Company of Hazelhurst.) The logging railroads referred to here were all laid down to bring out the last of the big hardwoods, mainly hemlock.

Engine No. 92 for John R. Davis Lumber Company approaches Phillips from Camp I over Lake Shore & Eastern tracks. This engine was probably built by Baldwin and purchased about 1890. It was used for spur and mainline work and sold for scrap in early 1900s.

The John R. Davis Lumber Company office staff, about 1890, at Phillips. Front, left, Mark Boulton and Pearl Bentley; middle, left, H.M. Crittendon, Thomas R. Rowlands, Earl Lace, John R. Davis and Ben W. Davis. Back, left, L.J. Ripley, Sam Monroe, E.J. Riordan, DeWitt Van Ostrand, and J.F. Roberts. Crittendon, who was from Chicago, is described in the Phillips *Times* on February 23, 1889, as "an expert accountant."

Head office of John R. Davis Lumber Company at Phillips. Stoves have heat conductors, and office staff work at roll top desks. Man at right is probably using Oliver typewriter. This building is presently occupied by Hardware Hank store operated by Art Johnson.

Log train enters Phillips, winter of 1908-09. Lake Duroy (right) is covered by ice. Engine No. 258 is pulling sixteen Russel cars of logs from camps on Elk River east of Phillips. Logs will probably be rolled on ice until spring. View looks north to sawmill built after fire of 1894, with store buildings at left on Lake Street.

Marker of State Historical Society of Wisconsin on Highway 13 overlooking Lake Duroy calls attention to fire of 1894.

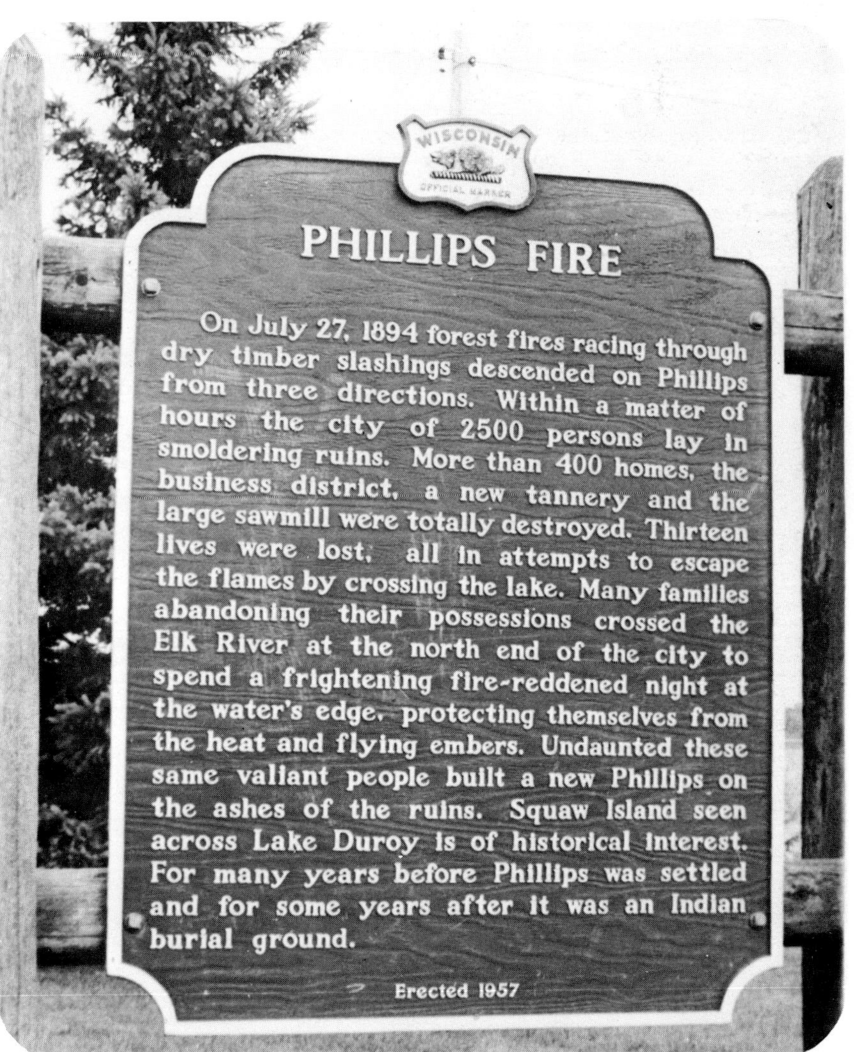

Above, camping party of families associated with the Phillips Lumber Company. Engine No. 92 was probably the one referred to locally as the "John R" which was used both on spurs and as switch engine. This picture was probably taken in summer of 1889 when the company was still being called Phillips Lumber Company, but was changed in July to John R. Davis Lumber Company. According to the Phillips *Times* under date of February 4, 1889, company cut 12,000,000 feet of timber in winter of 1888-89 from its "Sailor Creek timber." This picture was taken near Sailor Creek dam, and the campers, l. to r. are Tony Andrae, Ben W. Davis, Fred Rogers, Mrs. L.J. Ripley and Mr. Ripley, Mrs. Fred Rogers, next unknown, and next George Carter. Last four on right unknown.

Below, construction technique used in building track across swamp ground is shown here in scene from Lake Shore & Eastern Railroad, east of Phillips. Corduroy has been laid on strip cleared of timber and brush, and ties laid over corduroy. Rail lines like this could be ripped up and moved to new stand of timber without much difficulty.

Although from Bayfield, this Wisconsin Central No. 10 was photographed in work train service at Phillips, probably in early 1890s. Engine was built by Baldwin in July 1872, C/n 2859. The 4-4-0 had 17" cylinders, 60 3/4" drivers and weighed 70,000 pounds.

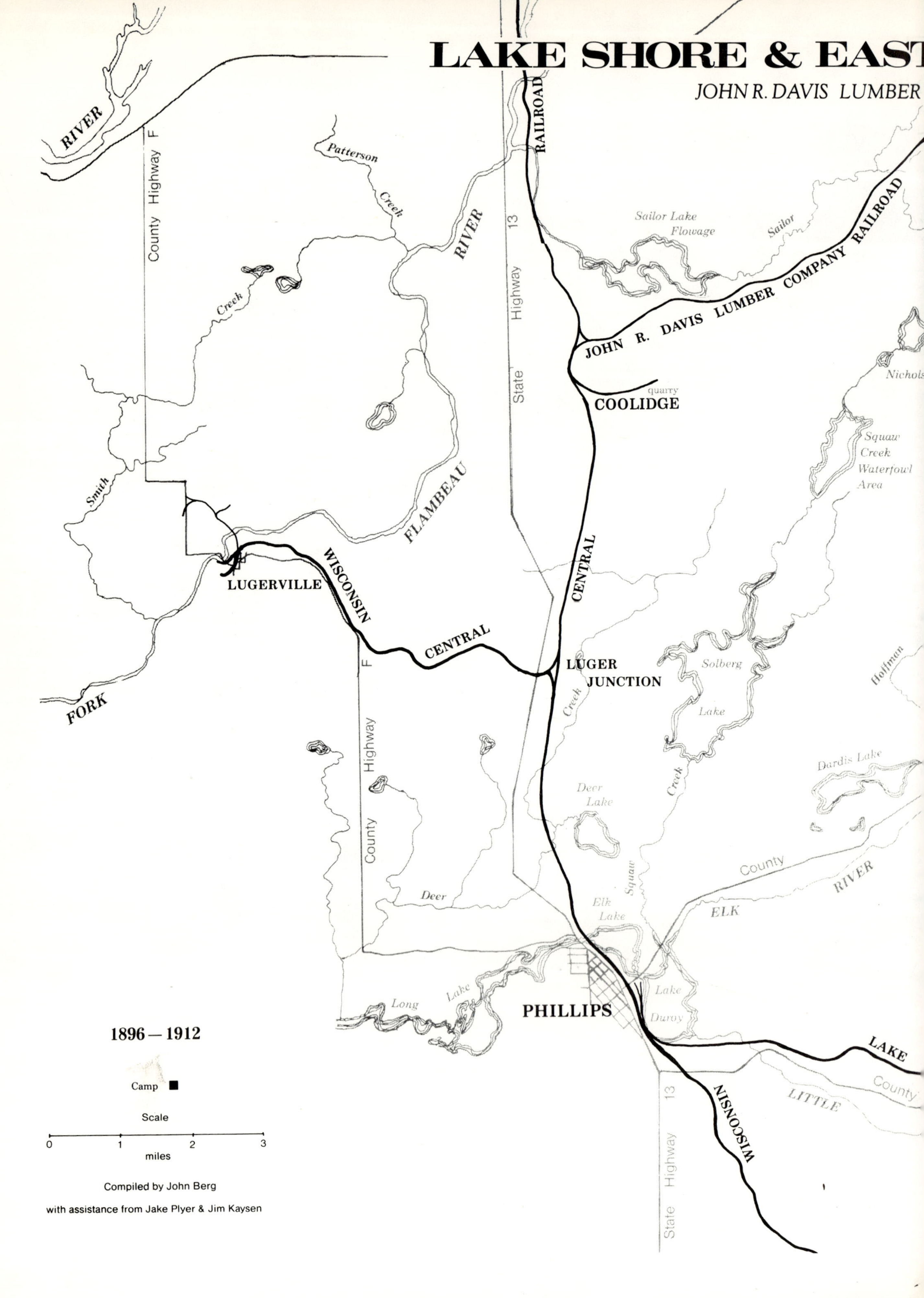

LAKE SHORE & EAST
JOHN R. DAVIS LUMBER
RIVER
County Highway F
Patterson
Creek
RIVER
State Highway 13
RAILROAD
Sailor Lake
Flowage
Sailor
JOHN R. DAVIS LUMBER COMPANY RAILROAD
Creek
quarry
COOLIDGE
Nichols
Squaw
Creek
Waterfowl
Area
Smith
FLAMBEAU
CENTRAL
LUGERVILLE
WISCONSIN
CENTRAL
LUGER
JUNCTION
Hoffman
FORK
Creek
Solberg
Lake
County Highway F
Creek
Dardis Lake
Deer
Lake
Deer
Squaw
County
RIVER
Elk
Lake
ELK
Long
Lake
PHILLIPS
Lake
Duroy
LAKE
County
LITTLE
WISCONSIN
State Highway 13
1896 — 1912
Camp
Scale
0
1
2
3
miles
Compiled by John Berg
with assistance from Jake Plyer & Jim Kaysen

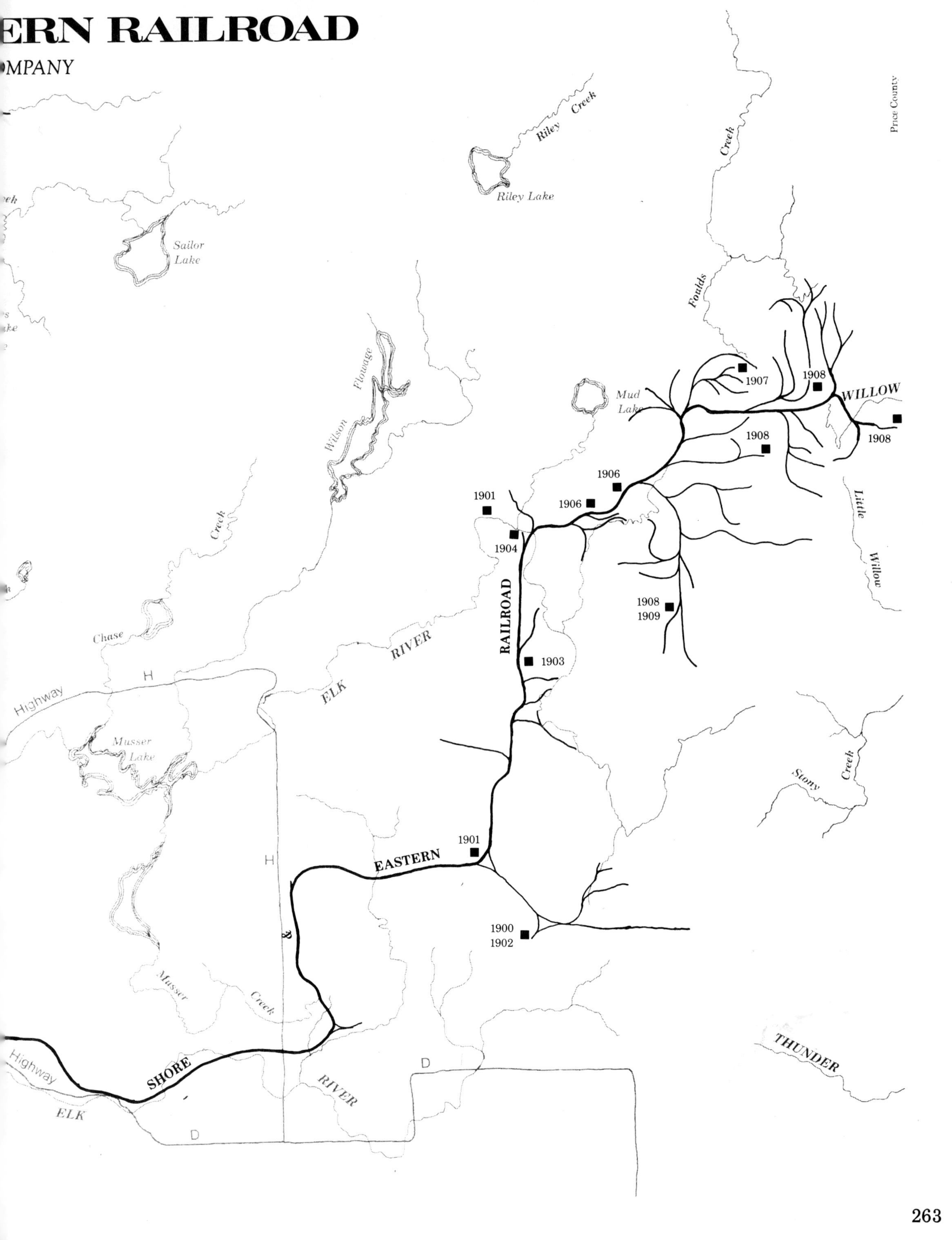
ERN RAILROAD
MPANY
Riley Creek
Riley Lake
Creek
Price County
Sailor Lake
Foulds
Wilson Flowage
Mud Lake
1907
1908
WILLOW
1908
1908
1906
1901
1906
1904
Creek
Little Willow
1908
1909
RAILROAD
Chase
RIVER
1903
H
Highway
ELK
Musser Lake
Stony Creek
1901
EASTERN
H
1900
1902
Musser
Creek
THUNDER
Highway
SHORE
D
RIVER
ELK
D

West Lumber Company mill at Lugerville on east bank of south fork of Flambeau River. Log cars are being unloaded at left into pond. Silo-like structure is slab burner. Below: planing mill at Lugerville located short distance east of sawmill.

Lowrie Lowe (in white shirt) mill superintendent at Lugerville mill, prepares to board Soo Line caboose No. 99001 at Luger Junction.

Soo Line ran special "banana run" from Phillips to Lugerville on Saturday night to bring home students attending high school in Phillips, and returning on Sunday night to Phillips.

Engine No. 218, a standby on the Lake Shore & Eastern railway, was eventually moved north to new sawmill town called Morse, north of Phillips. In picture above man on tender is Sam Anderson, man with oil can, Sanford Shell, and upfront, John Anderson.

Center: Engine No. 101, one of two used to haul logs from Camps 6 and 7 for West Lumber Company, to mill at Lugerville.

Left, Lowrie Lowe, at Lugerville Junction, waiting to board caboose. White flag on engine signifies extra train.

George Knoblock, engineer, stands by No. 102 for West Lumber Company.

West Lumber Company's 101 pumps water into boiler of steam jammer (left) used in loading operations on landings in Flambeau township, west of Lugerville, Price County.

Log train on Omaha line pulling into Couderay, Wisconsin. Watertank up front, mill and yards at right, slab burner at left, probably for Hines Lumber Company.

Wreck on Omaha branch at Donald east of Holcombe, about 1905.

Section crew at Holcombe, Wisconsin, about 1905. Seated in center is Alonzo Waldeck, depot agent, flanked on left (arms akimbo) by Eugene JuVette, and on right by Del Thatcher. Standing on handcar, left, Clarence Birch, next unknown, and next Guy Butler. Handcar was propelled by two or four men pumping handle up and down. Gasoline engine for handcars came decade later. Tank in back supplied train engines with water. At left is shed for tools and handcar.

The Melting Pot

The woods attracted a wide variety of men of different ethnic backgrounds. Somewhere, at the turn of the 20th Century in Big Bend township of Rusk County, there was a sawmill, probably not far from a landing just above the elbow of Big Bend on the Chippewa River. The federal census for 1900 gives the names of the crew working at the mill as well as their occupations, place of origin and date of birth. The variety of names suggests that a sawmill crew in the heart of the northern Pinery could also be a "melting pot."

The spellings used by the census enumerator leave much to be desired and in some cases, the name can not be read accurately. A question mark will caution the reader. Other names, for example, such as Peter Gilbon are, no doubt, misspelled. Gilbon was from Norway and the name probably should have been spelled Gulbrand or Gulbrandson.

The enumerator usually gives the first name in a list of associated names as "head" of family, and since Ole Anderson was the cook, he was considered "head of family" and his name appears first on the page. He was born in Norway. The others working around and in the mill follow:

Jacob Lacken [probably Loken] second cook, born in Norway.

Frederick Tragart, second cook, born in Wisconsin, parents from Germany.

Duncan McPhee, day laborer, born in Canada.

Martin J. Maher, well digger, born in Minnesota, parents from Ireland.

Richard Warieu (?), well digger, born in Minnesota, parents from Ireland.

James Duquay (?), carpenter, born in Canada.

Michael McLaughlin, carpenter, born in Minnesota.

Thomas Slatterup, locomotive engineer, born in Canada, parents from Ireland.

Eugene Gorman, locomotive fireman, born in Minnesota.

Hugh M. Lain, carpenter, born in Indiana, parents from New York state.

Samuel Whittaker, carpenter, born in Wisconsin, parents from Scotland.

Carl Stromberg, day laborer, born in Sweden.

John McDonald, teamster, born in Wisconsin, parents from Scotland.

Harry McMillen, mill foreman, born in Michigan, parents from Scotland.

Harry Austin, edgerman, born in Wisconsin.

Elmer E. Cole, laborer, born in Michigan.

James Fisher, laborer, born in Maine, parents from Canada.

Thomas Carpenter, laborer, born in Indiana.

Alphones Korm, laborer, born in Germany.

Norton P. Morgan, laborer, born in New York state.

Bert E. Culver, carpenter, born in Indiana, parents from England.

Julius Johnson, laborer, born in Norway.

Robert Tilton, millwright, born in Virginia, parents from England.

James Rierson, lumber piler, born in Wisconsin, parents from Norway.

James Argetsinger (?), lumber trader, born in New York state.

Peter Lofthus, lumber piler, born in Wisconsin, parents from Norway.

Henry Bloom, lumber piler, born in Russian-Poland.

Anton Pawala, lumber piler, born in Germany.

Henry Ramos (?), lumber piler, born in Germany.

Peter Gilbon, lumber piler, born in Norway.

Amund Hanson, lumber piler, born in Norway.

John Larson, blacksmith, born in Sweden.

Bert Smith, timekeeper, born in Wisconsin, parents from Indiana.

Joseph Byron, millwright, born in Wisconsin, parents from Canada.

Frank Bliss, head sawyer, born in New York state.

John Larson, laborer, born in Sweden. (There were two John Larsons in the crew but their ages were years apart.)

Fred Beaugard, teamster, born in Wisconsin, parents from Canada.

Harry McMillen, the mill foreman, was only twenty-five years old, an achiever of the first rank. The head sawyer, however, was over forty.

In the years 1885-1886 Thomas Meagher of Chippewa Falls operated a store near the Little Falls dam on the Chippewa. Perhaps he ran the

Behind beards and unshaven faces, it would be difficult to tell nationality of these men, whether Norwegian, Polish, Irish or Yankee. At table all were equal and all had big appetites. Picture taken in mess hall constructed on railway flat car for Hines Lumber Company spur to Camp 3, Hunter Lake, Sawyer County, winter of 1920-21.

wanigan located in the boat house. Whether in the store building or the boat house, he left behind a ledger containing his accounts for one year. Charles ("Kib") Ecker, who worked on the company farm at the dam for many years, came into possession of the ledger and after he died, his daughter Marian (Mrs. James Carroll) of Holcombe found it in her father's papers. It is a rare document, for here are the names of some of the men who drove the Chippewa River, built its dams and broke its jams. Although Meagher did not know how to spell the names of all his customers, he need not be faulted for it because many of them did not know how to spell their own names. Thus there is a Byron Gormly and Chas. Gormerly, almost surely members of the same family. But other names in the ledger leap out at us, such as Wm. H. England, one of the engineers who built the Little Falls dam, James Jardine, later a boss at the dam, John Hedrington, who was almost shot when he got in the line of fire between John F. Dietz and Sawyer County deputies in the Battle of Cameron dam in

1905, and Eugene JuVette, who was involved with Luke Lyons in sculpting the Indian Brave. Luke Lyons, the creator of the Brave, is not mentioned in the ledger since he had left the company.

Here, then, are the names of the men as they seem to be written in the ledger:

Lewis Vain
And. Murphy
James Harrison
Alex Laundry
Jos. Blair
Ole Eggan
John Green
Fred Hess
Frank Covill
Robt. Taylor
Thos. Marlow
O. Laduck
Geo. Dickson
Pet. Miller
James Jardine
Wm. Halems
Geo. Patrick
Martin Christenson
Mike Haley
James Demsey
Dan Kelly
Wm. Strong
Robt. Murry
Edw. Johnson
Mike Monoghan
Alfred Nassier
David Moore
Dan Duncan
Barney Dolan
Jos McGilvery
Joseph Cossett
David Savia
Frank Sabay

Jos. Russet
Mitchell Pelchu
Jim Martin
Rook Murray
Mart. Lyons
John McInnis
Edw. Carey
Wm. Campbell
Wm. Rooney
Dale Pitsch
John Cameron
John Haley
? McCoy
Roark Rasmussen
Jos Bushy
David Walker
John Mitchell
Neal McIntyre
James Croker
Andrew Allman
Hubert Bodin
Barney Dolan
Chs. Shaw
Phil Thornton
Leopold Duccoumin
John Courford
Stuart Mingo
Lewis Gatieu
Alfred Sutherland
Geo Simes
Jos. Sight
Wm. Barrett
Pat. McHugh

Angus McDonnell
John Grant
Albert Riordan
David Dussett
Christ Malon
John Wade
Melvin Albert
James Dolan
James Dempsey
John Haley
Thos. Carrell
Pat O'Neal
John Dumas
Luke Lyons
Jos Russett
Wm Haley
Chas. Gormerly
Francis Gigan
Henry Gannon
Patrick Burke
George Burgess
Joseph Vaugoux
Victor Moline
Louis Lemery
James McFarland
Charles Richard
Louis Debrot
Henry Gleason
Octave Lema
Wm. H. England
Louis Prue
Mike Quin
Randal Harison
Isreal Cuturia
Rudolph Pitsch
Thomas Marlow
Mike Ulticar
Wm. H. Hagan
David Campel
Mannin Pitsch

John McTeague
James Boyle
Rory McPhee
Lewis Sweeney
George Dickson
Finley McDonald
James Ryan
Joseph Fuch
Alphonse Gardin
Eugene JuVette
Otto Pitsch
Emil Moline
Pat McGouldrick
Dolay Pitsch
John Hedrington
Adolph Pitsch
Jos. Joco
G. Reslow
(Glode Russlow)
Joseph Joas
Charles Corbine
? Hamilton
Ed Jonson (Johnson)
Jos. Fox
Thomas Sulivan
Wm. Doner
Charles Mallory
James Mallory
James Burns
Bryon Gormly
J.S. Phillips
George Peterson
Wm. Besom
John Bessom
Jessie Tebart
John McCallum
James Carrol

Nathan A. Preston opened a photographic studio at 224½ Fourth Avenue in Eau Claire in the mid-1860s and in 1867 he was commissioned by *Harper's Weekly* to go to Chippewa Falls to take a picture of a big log jam above the city. Picture of jam, which appears elsewhere in this volume, seems to be only logging picture in Chippewa Valley to survive from 1860s. In self-portrait above, Preston and wife pose for a later generation to study them and consider their place in history. He sits with elbow on book, probably the Bible, and she modestly stands behind stuffed chair. Stereopticon slides lie in pile on table, one mounted in stereopticon. Top silk hat was posh headgear for gentlemen of the period. Mrs. Preston probably wears her favorite dress which is trimmed with velvet. Preston continued in business at Eau Claire until his death in 1904.

The Chippewa Trail

The Chippewa Trail north from Chippewa Falls into the Pinery may have begun as an Indian trail, a single path worn smooth by Indian poles dragging travois. After the white man came into the Pinery, probably this same trail was enlarged for wagons and sleighs. The use of the word "trail" for a road continued to be used after the turn of the 20th Century. For example, early automobile routes were referred to as trails, one of the more familiar: the Yellowstone. Another motor road was called the "Red Dog Trail." Until highways were numbered, it seemed to be the most comfortable to associate a road with a trail.

In the late autumn when the lumberjacks in the Chippewa Valley went north, they followed a vector into the Pinery called the "Chippewa Trail," which, in early spring and late fall, was heavily traveled. There was a need for some kind of stage service to be sure. One traveler in November 1873, passed sixty horse teams and about 200 men on foot, all moving north to the lumber camps.

Amos A. Stiles and Peter Lego, who operated a livery stable in Chippewa Falls, saw an opportunity to expand their business by running a stage line into the north as far as Big Bend on the Chippewa River, a distance of about forty-five kilometers as the crow flies, but fifty-five as the stage wheels ran. An advertisement in the Chippewa Falls *Herald* on May 3, 1878, announced that the company had "reduced fares of $2 from Chippewa Falls to Big Bend." The advertisement also said the stage ran every other day, that is, it took one day to reach Big Bend, and the second day to return. But this schedule was revised shortly when the route was extended to Trading Post near the forks of the Chippewa River, today in eastern Sawyer County.

There is, apparently, no engraving or photograph available of the Stiles stages. They were, most probably, a light wagon with canvas over the top and spring seats on the wagon box. After snowfall, the line suspended operating for the season.

One source believes that it took fourteen hours to reach Trading Post, but this, patently, is an exaggeration. No team of horses on a wagon of any kind could keep up a pace that fast over that many hours. There was also the uncertainty of the weather, or a tree in the road, or an unexpected mud hole, or a horse that threw a shoe.

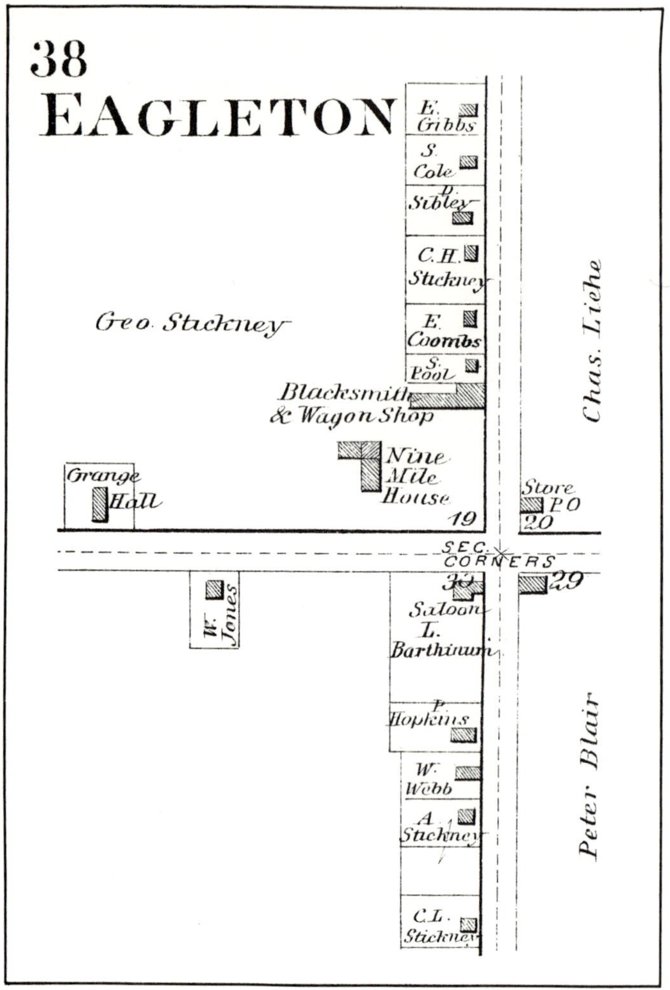

Detail of plat of Chippewa County for 1888 shows location of Nine Mile House and community called Eagleton north of Chippewa Falls.

The *History of Northern Wisconsin*, published in 1881, includes a chapter on Chippewa County and among the interesting bits of information is a paragraph on the stopping places passed by the Stiles stage line, to wit:

> "The stopping places on the road are as follows: Nine Mile House; Twelve Mile House; ten miles beyond is Campbell's; three miles is Larrabee House; one to the Lake House; nine to Big Bend; one beyond is Allen's; three more to Oak Grove; twelve to Johnson's; four to Pinkham's; four to Murray's; twelve to Hermon (sic) House; six to the Hall House; three to M. Sarrow's; four to West Bend; and four to the Trading Post . . .

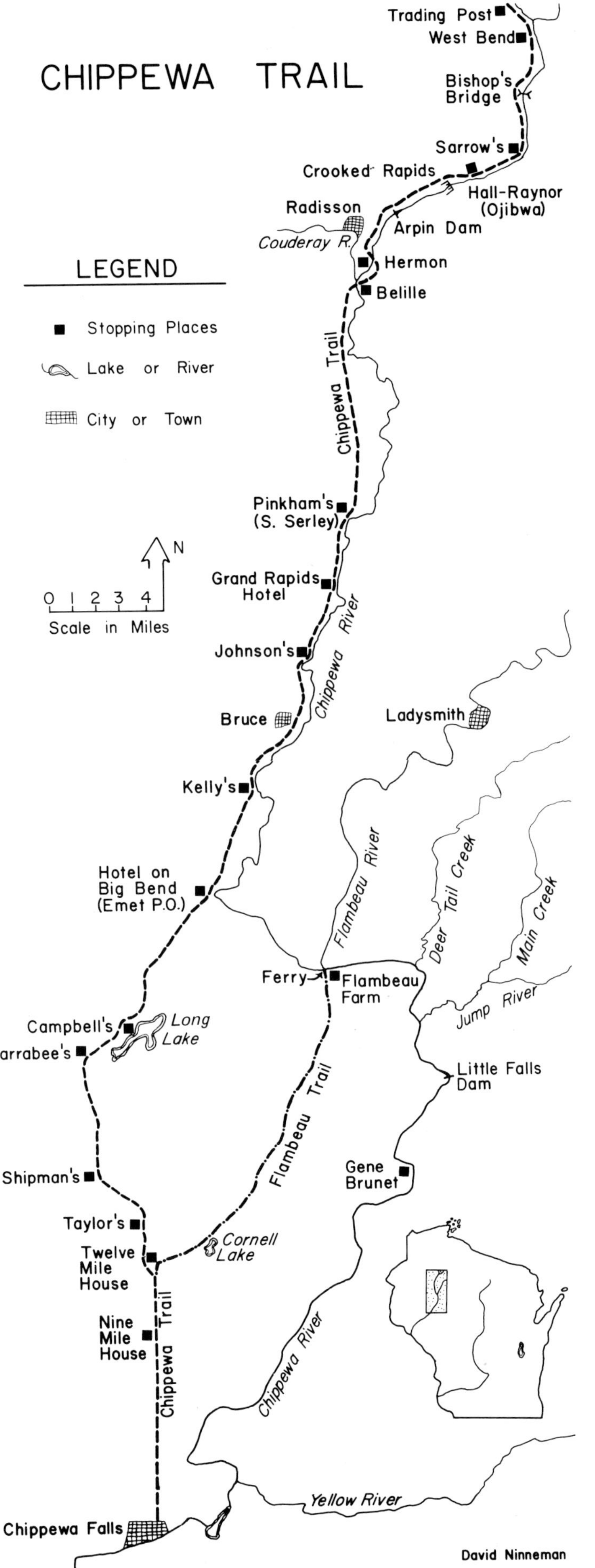
CHIPPEWA TRAIL
Trading Post
West Bend
Bishop's Bridge
Sarrow's
Crooked Rapids
Hall-Raynor (Ojibwa)
Radisson
Arpin Dam
Couderay R.
Hermon
Belille
LEGEND
Stopping Places
Lake or River
City or Town
Chippewa Trail
Pinkham's (S. Serley)
N
Grand Rapids Hotel
0 1 2 3 4
Scale in Miles
Chippewa River
Johnson's
Bruce
Ladysmith
Kelly's
Hotel on Big Bend (Emet P.O.)
Flambeau River
Deer Tail Creek
Main Creek
Ferry
Flambeau Farm
Jump River
Campbell's
Long Lake
Larrabee's
Little Falls Dam
Flambeau Trail
Gene Brunet
Shipman's
Taylor's
Cornell Lake
Twelve Mile House
Nine Mile House
Chippewa Trail
Chippewa River
Yellow River
Chippewa Falls
David Ninneman

Most texts drop the article *the* in referring to Trading Post.

The itinerary carried the stage line through modern Rusk County into modern Sawyer County, both of which were set off from Chippewa County in the 1880s. (Rusk County was first called Gates.)

Most of the stopping places listed in the above are identified on the 1888 plat of Chippewa County as far north as the south line of Sawyer County. But there were some changes, too.

On the plat book the stopping places are identified as "hotels" or as a "house," although in common parlance they were all called stopping places, i.e. a place to stop over night. Anyone could establish a stopping place, for it was merely a place for a person to stay over night, and equally important, it was a place where a teamster could find a stall to bed down the horses for the night. Quite often, the teamster slept in the barn, or "hovel" as it was called.

Ed Campbell's Lake House, much remodeled, on Long Lake in Chippewa County.

Historical marker erected by Tony Wise on site where Charles Belille (prounced Bee-lil) settled on Chippewa River in late 1830s. Belille was French-Canadian and first white settler in modern Sawyer County.

The route taken by the Stiles stages followed very closely the route today followed by modern Highway 40, except for a few turns and bends, all the way to the junction of the Couderay and Chippewa Rivers (south of Radisson). Here the trail turned northeast after crossing the Chippewa River to the east bank, and a short distance farther on, recrossing it to the west bank.

Early roads in many instances followed the section lines, but in the lake and river country, a road followed the lines of least resistance, avoiding swamps and sloughs, and always seeking out the best place to ford a river.

One of the best known stopping places on the Chippewa Trail lay north of Chippewa Falls at "Nine Mile House," so-called because it was about nine miles north of the city. There was a room for dancing here, a fact confirmed by the keeper of the daily journal at the Little Falls Dam who wrote on July 4th: "Dance tonight at the 9 Mile House but too far for us to go." A dance floor of the period could be a living room with furniture pushed back to the wall, and a fiddler in the corner belting out old camp songs from the Civil War.

George Stickney owned the forty on which Nine Mile House stood, and Stickney and his wife were the original owners. Mrs. Stickney is usually given credit for naming the community "Eagleton," a name which survives into the present.

Continuing due north, the next stopping place was Twelve Mile House, actually less than three miles north of Nine Mile House. It lay within the angle where the Chippewa Trail bent northwest, and another trail, called the Flambeau Trail, which went north and east. P. Pinkham paid taxes on the two forties on which Twelve Mile House stood in 1888.

As the map shows, the trail from Twelve Mile House went northwest and then bent west into Bloomer township for about two kilometers, and then turned due north. Less than a kilometer into the town of Sampson, in the southeast corner of Section 35, was I. H. Shipman's stopping place. J.E. Campbell owned the two forties east of Shipman, although he did not live there. Campbell, in 1888, had a stopping place farther north on Long Lake.

The next stopping place was "Larrabee House" located on a small lake today known as Larrabee Lake west of Long Lake. But it was only a short distance up the road to "Lake House," the stopping place operated in 1888 by J.E. Campbell & Son. It stood directly on the range line between Section 13 in Sampson township, and Section 18 in Birch Creek township. Thus, there is strong reason to believe that Larrabee House was here first and that Lake House was erected to take advantage of a special circumstance, namely, a sawmill which is identified on the 1888 plat as "Snow's," located on the west shore of Long Lake, with Lake House within easy hearing distance of the sawmill whistle. This was, no doubt, a mill with a small crew which Lake House could cater to without hurting its transient business.

Lake House may have been founded by a man named Nuff, and taken over by Edward Campbell, a former steamboat captain from Durand, and father of J.E. Campbell, but it is quite possible that Captain Campbell ran Shipman's stopping place first, and that either he or his son moved north to take over, or to build, Lake House. The 1888 plat indicates no building here other than a farmhouse.

Stopping place of Al Raynor as it looked in 1924 with old bridge across Chippewa River at right. New bridge was built at left. Log walls in house are still intact but outside has been reboarded and interior remodeled to conserve energy. House is oldest standing in Sawyer County and is on National Register of Historic Places.

It was nearly fifteen kilometers from Lake House to a stopping place on Big Bend, a focal point on the Chippewa River in lower Rusk County. An early map of the pine regions dated 1883 identifies only a "hotel" at Big Bend, no name, and the "Emmet P.O.," a misspelling for Emet, a post office established in 1878. It was later moved a short distance north and in 1892 changed to "Island Lake," a post office discontinued in 1933.

The 1888 plat identifies only a school house

Thaddeus Thayer's house which probably replaced earlier log shanty, known as Trading Post, was turn-around point for stages out of Chippewa Falls. Trading Post stood at mouth of Pokegema Creek on west fork of Chippewa River, but site today covered by Chippewa flowage.

near Big Bend. In the 1880s the Chippewa River & Menomonie Railway hauled thousands of logs to a landing just around the Bend to the north, and the Chippewa Trail ran past what may have been the depot.

Some passengers on the Stiles stages got off at Big Bend to use the shanks mare along the west bank of the river southwesterly to Flambeau Farm lying opposite the mouth of the Flambeau River, headquarters of operations in the north for the Daniel Shaw Lumber Company.

From Big Bend the Chippewa Trail ran northeast, crossing Soft Maple River and Potato Creek, while clinging to the west bank of the river. *The History of Northern Wisconsin* refers to a stopping place one mile north of Big Bend as "Allen's," but the 1888 plat fails to include it which suggests that Allen had gone out of business. The same source mentions "Oak Grove" as a stopping place three miles beyond, that is, north of Allen's. Actually, it was closer to two miles north where the 1888 plat identifies a post office called "Emet" on Section 1 of Big Bend township in Rusk County, as already mentioned. The post office in 1888 was located on a farm operated by B.F. Brainard and it is quite possible it was situated in Brainard's house.

After crossing Potato Creek, the Chippewa Trail bent around the east shore of Amecoy Lake and here, the plat of 1888 identifies a "hotel." The land on which it was situated is listed under the name of James Kelly, and the business directory of the plat book lists Kelly as a farmer and stock grower, justice of the peace, and keeper of "Kelly Hotel."

From Kelly's Hotel, the Chippewa Trail clung to the west bank of the river for about three kilometers and then swung north to Johnson's stopping place. *The History of Northern Wisconsin* lists Johnson's "twelve miles north of Big Bend." This was S. (for Sam) J. Johnson who lived about three kilometers north of the present village of Bruce. The business directory of 1888 plat gives the name of S.J. Johnson as the proprieter, but the plat map shows that although someone was living on the Johnson quarter section, there was no hotel here. It was, however, one of the first on the Chippewa Trail. It may have been built in the 1860s at a point where there was once a ferry across the Chippewa River. Oral tradition holds that when the Soo Line Railroad built here in 1884, Johnson moved to the newly-platted village of Bruce to open the "Johnson Railway Hotel."

Less than two kilometers north of the one-time Johnson place, the 1888 plat identifies a "hotel" on the Chippewa Trail, and the forty on which the

hotel stood is attributed to J.T. Thrasher.

About two kilometers north of Thrasher's stood "Grand Rapids Hotel," built by Cy Pinkham, probably in 1871. It seems that he was related to the Pinkham mentioned earlier. The plat of 1888 lists B.F. Brainard as the owner of the forty on which Grand Rapids Hotel stood.

The Pinkham establishment became an important rendezvous for logging operations on the upper Chippewa River and tributaries such as the Couderay and Weirgor. There was a company store,

Picture, allegedly dated 1905, purports to be first house in Kennedy, a community west of Park Falls, Wisconsin (now a ghost town). Log shanty shown here was typical of homes built on the cut-over lands and slashings left behind by the lumbermen. Pioneer families bought land from the lumber companies or from Chicago developers, but great many lost their homes for want of capital to buy machinery and cattle, or because soil was unsuited to farming. In World War I draft dodgers bought slashings and began "farming" but fled back to cities after war was over. Most small farms in north were absorbed by second generation of farmers who had more agricultural expertise and enough money to pay down on farm without assuming burdensome mortgage.

or wanigan here, two blacksmith shops, sleeping shanties for transient crews as well as permanent party, and a barn that accommodated scores of horses. The 1888 plat shows that there was also a ferry here. Just when Pinkham sold out is uncertain, but D.W. Blackburn is known to have operated it before it burned in the late 1880s.

The History of Northern Wisconsin next lists "Murry's" stopping place four miles north of Grand Rapids Hotel. The Chippewa Trail north from Grand Rapids Hotel followed the river a short distance and, where the Chippewa bends east, the trail continued north to Weirgor ford. But the name of Murry does not appear on the 1888 plat. Instead, the two forties where Murry's presumably stood are listed under "S. Serley." This was Sever Serley, a Norse immigrant who learned to cook in the logging woods in the 1870s, and about 1880, bought the Murry stopping place which was reputed to be a three-story building. This burned in 1888 and was rebuilt of frame construction. Business continued until 1890 or so when Serley decided there was more money in logging than cooking and became a jobber. But in 1900, the federal census lists him as a "farmer." Other sources confirm that Serley had given up logging and returned to farming.

Nearly twenty kilometers up the Chippewa River from Serley's place lay Hermon House which was probably operated by Joseph Hermon (also written as Herman) and located about three kilometers south of what came to be the Arpin dam. In fact, the construction crew on the dam occupied some of the abandoned buildings, but today only the footings are visible.

The modern plat of Sawyer County, Weirgor township, shows a cemetery along the north line of Section 2. This is the old Catholic cemetery overlooking the Chippewa River on the left (east) bank, and a stone's throw away, stood the homestead of Charlie Belille who is considered to be the first white settler in the present limits of Sawyer County. Belille, a Frenchman, was not conducting a stopping place, but travelers sometimes took refuge here in a storm. It was not the most comfortable place because there were two many people around, mostly children. Belille was married to a Chippewa woman who is remembered in English texts as Esther Crane, and he later took to wife two more Chippewa women and sired a family of more than twenty children.

Phillis Sander, a local historian at Exeland, writes that Belille lived at La Pointe for three years and, probably in 1835 or 1836, came south with a band of Chippewas to the lake country around the present Indian Reservation. From there he canoed out of Lac Court Oreilles down the Couderay River to the Chippewa where he found a place he liked and built a log shanty.

Belille had a trap line and traded with the Indians When Anthony Hayward stopped at his place one night in the late 1870s, having walked up from Chippewa Falls on the ice, he told Belille that logging operations would soon begin in this part of the country, and Belille hurried to clear more land to grow more vegetables and hay for the logging camps. For many years he also provided a portage service around Belille Falls, a place of uncertain location today. A small historical marker, erected by Tony Wise, calls attention to the site where the Belille homestead stood.

When the Stiles stages approached this area, they crossed the Chippewa River east of Windfall Lake, or, near Belille's place, to gain the east bank, and from here the stage went on to Hermon House on the west bank which required another river crossing.

But here the Chippewa is not deep and fording the river was probably no problem. After crossing to the west bank to reach Hermon's place, the Chippewa Trail continued to follow the west bank all the way to Trading Post, terminus of the trail.

Nine kilometers up the river from Hermon House stood the Hall stopping place which, incidentally, is still standing where it was built on the north side of the bridge at Ojibwa at what is today the intersection of Highway 27 and 70. It is believed that the Hall stopping place was built in two periods, first as a one-room log house, and the second, when a big wing of two rooms on the ground floor and more rooms on the second floor were added. The first log house was probably built by Charles C. and Sarah F. Drew who had acquired the lot on which the building stands from Eric

McArthur on February 13, 1872. The Hall brothers, Alfred and Samuel, appear to have purchased a majority interest in the place from Drew and operated it as a stopping place until they sold out to Anthony J. Hayward and Warren E. McCord in 1878. Moving into the building on March 12, 1878 were Alfred and Theresa Raynor, and on June 7, 1884, they purchased the property from the last owners who may have been McCord and the Laird, Norton Company of Winona, Minnesota.

"Al" Raynor was born in Canada and moved to Maine and from there entered service in the Union Army during the Civil War. He came west in 1878 and took over the operation of the former Hall stopping place. As a veteran of the Civil War, he also knew a thing or two about small arms and on one occasion, according to oral tradition, he got into an argument with one Archie Moore on the opposite side of the river. Threats were exchanged across the river, and Raynor became so angry he went into the house to get his rifle and took a shot at Moore, but his aim was off and no serious damage was done. After 1895, there was little call for a stopping place here. The timber was nearly all gone and Raynor had to take up farming. He was a member of the County Board from Couderay township in the early 1900s. At the Hall stopping, in its heyday, one witness recalls, "you had to step high at night after the stoppers retired, as 100 men often slept on the floor."

The old Hall stopping place, recently refurbished with new siding of western pine and new interior panelling to conserve heat, is probably the oldest building in Sawyer County. Through the efforts of Eldon Marple, it has been placed on the National Register of historic places.

The *History of Northern Wisconsin* refers to the next two stopping places farther up the river from Hall's as "M. Sarrow's" and "West Bend." Little is known about Sarrow's. It apparently did not remain in business long. The place stood a short distance below what came to be called Bishop's Bridge.

The next stopping place may have been called "West Bend" at one time, but it was later known as Paul Lessards, and longest as "Little Tommy's" place after Thomas Manville, and apparently "Little Tommy" closed the place after the logging era ended.

The last stopping place and turn-around for the Stiles stages was Trading Post operated by Thaddeus Thayer, also one of the first white settlers in Sawyer County. He built a log shanty at the mouth of Pokegema Creek overlooking a bend in the Chippewa River, probably in 1865. He was squatting on Indian reservation land but the Indians apparently welcomed a man like Thayer who came with his part Chippewa wife, Mary Anibish Berg, from Chippewa Falls. But this stopping place is only a memory today, bound to oral report, buried forever at the bottom of the Chippewa flowage which covered it when the Winter dam was completed in 1923.

The 1880s and 1890s marked the peak years in the pine harvest on the upper Chippewa and east and west forks. Considerable timber was also cut by the Indians on the Reservation, or leased to jobbers, and thousands of logs were rolled into the Couderay River, or at "Russell's Landing" located about three kilometers northeast of Exeland.

The Flambeau Trail, also served by Stiles stages, was not noted for its stopping places. A teamster could, by pushing his horses, actually reach Flambeau Farm from Chippewa Falls in one day. It is fairly certain, too, that many lumberjacks found a place to rest over night in the house or barn of Alex Pinegard who operated a sawmill on a small lake north of Cornell Lake.

Flambeau Farm lay opposite the mouth of the Flambeau River on the Chippewa and served as a staging area for crews of the Daniel Shaw Lumber Company of Eau Claire. There was a ferry on the Chippewa directly west of the mouth of the Flambeau, and river crews used the right or west bank of the Flambeau at least as far as the junction of the north and south forks. The hotel at Flambeau Farm catered not only to the lumberjacks but to salesmen and fur traders. Daily necessities were sold in the wanigan, and at the height of the pine operations in the early 1890s, there were a number of families of mixed whites and Indians living in the vicinity who did the field work and took care of the chores.

Letter from Irvine to Weyerhaeuser

F. Weyerhaeuser, President
O. H. Ingram, Vice President

E. W. Culver, Manager
Wm. Irvine, Secretary

Chippewa Lumber & Boom Company

Chippewa Falls, Wis. June 26 1882

Mr. F. Weyerhaeuser Pt.
Wellesley Mass.

Dear Sir:

I arrived here 23d inst. - find all in good shape mill running well - cutting altogether on Yellow River stock - We had a good flood from Little Falls Friday. and fair driving water is reported on most of the streams - regarding which you doubtless have been advised by Mr Barnett.

Our cut to Saturday night - 24th inst. is 19.647.272 ft. (rafted lumber) In accordance with your orders - night work ceases July 1st. which we all think will be attended with good results.

We have sold one raft (1.750.M) to the Eau Claire Lumber Co. St Louis - upon the terms talked. i.e: 9.15 11.15 & 15.15 at Reeds. with running added - making the price in Alton though. 10.25 12.25 & 16.25 with the proviso that in case the average of our sales - from this date - to the end of the season, shall be less than the above prices - we are to refund the difference - and in case the average is more than above. we are to receive the difference.

(Letter continued from page opposite)

We have had a call today from Mess. Ingram & Dulaney and have sold them 3,000,000 ft to be delivered at Winona immediately upon the same basis as sale to the Eau Claire Lbr. Co.

These transactions are understood by the buyers to be *straight sales*, and not for our accomodation. We advising them that it was not necessary for them to take the lumber unless they needed it. In each case the parties wanted the lumber and would have bought it in any count.

We now have sold twelve million feet of new lumber and ten of old. I have written Schulenburg & Boeckeler Lumber Co. Saint Louis, offering them a raft on same basis as sale to Eau Claire Lumber Co. or at a fixed price in Alton Slough= $11, $13 & $17. Will probably hear from them tomorrow, and if they accept will send them a raft which is now under way. We will make delivery to the Eau Claire Lumber Co. about next Monday.

The situation of our lumber is about as follows: (In round numbers)

Sold and delivered [Board Feet]	7,000,000
Laid up in Boston Bay	7,000,000
No. 8 underway, for Eau Claire Lbr. Co.	1,750,000
No. 9 do probably will go to S. & B. Lbr. Co.	2,000,000
At Reads Landing & in transit on Chippewa	2,000,000
	19,750,000
Manufactured to 24th	19,647,272

The lumber at Reads & in transit (2,000,000) and 1,000,000 of our cut this week will go to the Empire Lbr. Co. at Winona, as stated.

Our financial condition is as follows	
1st Natl. [bank] Chippewa Falls	$22,479.07
Union Nat. Chicago	10,511.84
	32,990.91
Bills receivable (A.S.M. & Co.) upon which we can realize if necessary	18,877.07
	$51.867.98

In addition to the above, we hold your checks on Rock Island $10,000 and we will realize from our sales to Eau Claire & Empire Lumber Co. within ten days or two weeks at least $75,000. Our notes which were paid by your drafts (at Union Nat. Chicago) are all in our possession and cancelled.

Herein find a/c sales to date—30/32/36—No. 1 No. 6 No. 2 & part of No. 7.

Hope you are having a pleasant trip and that yourself and family are well.

Very respectfully
Wm. Irvine

P.S. Don't hurry home. Everything is going all right.

First up the Dore Flambeau

In a chapter on "Recollections of the Early Days in Price County," which appears in a promotional farm land booklet, published about 1915, William Seeburger describes his ascent of the Flambeau and the forks of the Flambeau in the late fall of 1872. After following the life of a lumberjack for more than twenty-five years, he retired to Phillips, Wisconsin, and ran a saloon for a number of years and later a confectionary store. He was elected mayor of the city twice and held office longer than any candidate before or since. He died in 1931 at the age of seventy-nine.

In his account, Seeburger mentions "Dukameaux Rapids." This is probably a French rendition of a name for a rapids which once lay a short distance south of Ladysmith on the Flambeau. The rapids disappeared during construction of the Port Arthur dam and bridge, and demolition of the bridge some years later. The name Dukameaux appears in English texts under various spellings, e.g. Leopold Duccomin *had a charge account at a store in Little Falls in 1886-87.*

The Seeburger story, here slightly abridged, follows.

Early in the spring of 1872 I worked on Black River for a company of lumber men then known as the Black River Improvement and Log Driving Company, and continued in the said company's employ until August first of that year. At the time spoken of a J.C. Hewett was the superintendent for the company. At the close of the driving season Hewett came and told me that he was going to start a ranch with one Eph. L. Hackett on the Flambeau river, and asked me if I would like to go up there and work with them. I told him that I would consider the proposition and let him know later.

I finally concluded to go and on November 9th, 1872 Charles H. Roser and myself left La Crosse for Eau Claire and thence by stage to Chippewa Falls. Here we met Eph. L. Hackett, and on the 10th day of that month we staged it to what was then known as the Flambeau Farm, on the Chippewa, and at the mouth of Flambeau River. This farm, which became noted in after years as a starting point for a hike through the dense woods to the up river lumber camps, was owned by Daniel Shaw of Eau Claire. Here we stopped overnight.

William Seeburger

On the morning of November 11th we loaded three canoes with supplies, and of which one canoe was loaded with iron, such as was needed in the far distant camp. When all loaded Hackett assigned me to the position of bowsman to the boat loaded with iron, with one Jack Thacher as steersman, and Charles H. Roser amid-ship. Mr. Hackett took the second boat, with a Chippewa Indian, and the third boat was manned by two Indians. In this manner we started up the Flambeau River in the Fall of 1872. Our destination was what has since been called "Hackett's Farm," at that time the farthest up river lumber camp on the Chippewa waters.

The Flambeau is a rather slow river for about six miles up from its mouth. Then comes small rapids and swift water. Every thing went along well until we came to these rapids; then it did not go quite so nice. But, by using a little head work, and a greater amount of main strength and pure awkwardness we scratched along, got over these rapids and went ashore for rest and a "bailout." When rested we started on again and worked our way up river without much difficulty until we reached what was known as Dukameaux' Rapids. Here the Daniel Shaw Lumber Company had a logging camp, and here we took dinner. After dinner we started to pole up the rapids, and then the fun commenced. We would get a little ways when we would bump up against a rock and the boat would take a swing around in the current in spite of all we could do. Finally, when we got the boat right end foremost I made up my mind that the boat was not going to take any more swings. So, I put away the pole, jumped into the river, caught the nose of the boat and led it up over Dukameaux rapids. Then we worked up the river a couple of miles and went ashore to camp for the night.

Eph. Hackett acted as cook and made some tea and fried some pork. We had a little bread that we had got at Shaw's camp. While Hackett was cooking, the rest of us picked boughs for our beds. We had six blankets for seven of us. The two Indians had one bed, Jack Thatcher and I another; this left Hackett, Roser and an Indian for the third bed. Hackett told Roser to get in the middle and he and the Indian would take the outside to keep him warm. The next morning Roser was about squeezed to death by the strenuous efforts of Hackett and the Indian to keep under the blanket. The next morning Hackett got breakfast and we had the same menu that we had for supper the night before—excepting that we had no bread.

On the morning of November 12th, 1872 our boats were all frozen fast in the ice and we had to break our way out to the channel. That morning the river was running full of anchor ice and you can imagine what a task we had to pull our boats up stream. About noon we arrived at what was known as the Bruno Vinette Farm, then the fartherest camp up the river. This camp was about one mile up the Flambeau river from where Ladysmith now stands. Vinette's farm in the fall of 1872 was the head of navigation and we therefore pulled out our boats at this point and stored our stuff, which was afterward toted to camp.

On the morning of November 13th, 1872 we started out on foot for what is now known as Hackett's Farm, a distance of 22 miles on a newly cut tote road. This trip from camp to Hackett's Farm took three days. We had with us a team of four oxen, which had been sent to meet us from the farm, and all the four ox team could haul was about a thousand pounds of supplies. I remember that two years later Charles Biladreaux started from the Bruno Vinette farm with a four-ox load and when he got to Hackett's all he had on was one barrel of salt pork and a hash machine. He was on the road for three days with this load yet when he got into camp it was midnight. We were put to work the next day on building the camps, stables, blacksmith shop and store house.

About December 1, 1872, a crew was started out to cut a tote road up the North Fork of the Flambeau river, to section 13, town 38, range 3 west, a distance of about thirteen miles. Teams with supplies were started right after us. By this time there was about a foot of snow on the ground and the weather was very cold. The first night we spent on this work I remember well. We camped at the mouth of Connor's Creek. We had to shovel snow to get down to the ground and leaves for a bed. In those days we had the small shelter or shed tents. We would pitch these opposite each other and build a log or brush fire between them. While we had a good fire it would be comfortable; at least we thought so those days; but, we are afraid it would not be considered very comfortable now to the modern "lumber jack" boarding and sleeping in the John R. Davis' palace lumber camps on wheels.

By the second day after our arrival we had our tents comfortable, brush stalls built for the cattle and wood cut for the cook. C. H. Roser was the cook.

The next morning after we arrived at this camp

Interior of camp office which was usually occupied by bookkeeper or foreman, or, in this case, by the boss, Elmer Olson of Cadott who ran Olson Lumber Company. When this picture was taken, he was probably logging in Price County. Office like this was also called a "wanigan" since daily necessities could be purchased. Tobacco available seems to be mostly Peerless and Standard, both popular brands which could be chewed or smoked, but Velvet (lower right) was only for pipe or rolling cigarettes. Calendar on wall dated January 1922.

was a bitter cold one,—and right here was the first time I ever saw a cook getting breakfast with his overcoat on, a large sash tied around his face and a pair of big woolen mittens on his hands to keep from freezing. But, Roser was "going some" in those days and, in fact we all felt good and enjoyed our rough life.

For the first few days our whole work and aim in life was to build the camps as quick as possible. I think it was the 26th day of December when we moved into the new camps. In those days it was more difficult and took longer to build a camp than now-a-days. We had no lumber or shingles and every thing was worked out with the ax and adz. If we had an auger, saw and hammer we considered ourselves lucky indeed. We had bare nails enough to fasten the shakes on the roof of the cook shanty, and the rest of the work was put together with wooden pins or wedges.

It was about the first of January, 1873, when we finished building camp. In those days when the camps were built the boys were ready to commence logging, and as the timber was on the banks we did not have to stop to cut roads. That became necessary only as we worked back from the river.

This camp was 85 miles from Chippewa Falls, and 50 miles above where any other camp had been built on the river. It took seven days to make the trip to Chippewa Falls for supplies. The hay was all cut on wild meadows. The main staff of life for the men was salt pork and beans, blackstrap molasses and dried apples; flour and salt and tea. No sugar, fresh meat, or vegetables.

We banked 1,300,000 feet of pine logs from this camp that winter and broke camp on March 23rd, when the crew were sent down to the farm camp on the South Fork of the Flambeau River on the SE¼ of Section 5, Town 36, Range 3 West. Here the men were retained for the spring drive, and put to work building batteaux and driving tools. In those days all such tools were made in camp.

On May 3rd, 1873 the first drive started that was ever run on the North Fork of the Flambeau River. This drive was taken out successfully to the mouth of the Flambeau. Then a few of the men were sent back to Hackett's farm to begin the work of clearing land. This went on for the summer of 1873, and until fall. Then the firm of Hackett & Hewett dissolved and Hackett continued the logging business and located a camp on Section 13, Town 37 North, of Range 3, West. This was the first camp ever built on the South Fork of the Flambeau,—or Dore Flambeau as it was often called.

In the spring of 1874, the first dam was built on the South Fork of the Flambeau.

On the 26th day of October, 1873 the crew, with teams, cut and cleared a tote road on the Northwest side of the South Fork of the Flambeau to the camp location above mentioned. There was six inches of snow on the ground at that date. On the 15th day of November we commenced to haul logs on sleighs, and continued until the 7th day of April 1874. That spring after camp broke Allen Jackson took five men, a yoke of cattle and some supplies, went up the river on the ice and built a camp on the SW¼ of Section 10, Town 37, North of Range 2 West, and commenced to clear up a farm which is now known as the McKinley farm.

In the fall of 1874 the Wisconsin Central Ry. had its track laid as far as Worcester, (101) or Wolverine as it was better known in those days.

That same fall the late W.T. Price took a contract from the Mississippi [River] Logging Company to put in, or bank, 100,000,000 feet of pine logs. He located two camps. His headquarters camp was located on the SE¼ of Sec. 9, town 37, N.R. 2 West. The other camp was on the SW¼ of Sec. 17 in the same town. That same fall a tote road was cut from Worcester, west to mouth of the Elk river for a winter road. Also, one from Worcester to Elk Lake, where Phillips now stands. Supplies were hauled from Worcester to Elk Lake and thence boated down the Elk river and also down the South Fork or Dore Flambeau to Hackett's farm. During the winter the supplies were toted from Worcester over the winter tote road as far as Jackson's and Hackett's farms.

In the fall of 1875 a road was cut direct from Hackett's farm to Worcester, mainly following the line between townships 36 and 37. That fall, W.T. Price located his first camp on the Elk River, on Section 13, Town 37 North, Range 1 West: Another on Section 8, Town 37 North, Range 1 West, and also a camp on Section 13, Town 37, Range 2 West.

In the spring of 1876 W. T. Price attempted the first drive ever made on the Elk River, but it was not successful and was "hung up." In the spring of 1877 A. B. McDonald started the first dam to be erected on the Elk River, locating it on Section 11, Town 37, North, Range 2 West. It was built solely for log driving purposes. In the fall of 1879 and winter of 1880 Mr. McDonald built the second dam, known as "Job's Dam," located at the foot of Long Lake, on the SE¼ of Section 14, Township 37 North, Range 1 West.

Referring back to incidents of the year 1874:—

That year, 1874, Eph. Hackett, the pioneer logger of the Flambeau and after whom one of the towns of Price county has been named, formed a partnership with David Law and M. E. Mosher of La Crosse and the company thus formed purchased quite a large tract of pine timbered lands in Township 37 North, Range 3, West, and in Township 38 North, of Range 2 West, and commenced to log quite extensively. In 1876 E. L. Hackett retired from logging while Mosher and Law continued on until the summer of 1882, when David Law sold his interest to M. E. Mosher. Mosher continued logging until the spring of 1886, when he had all his pine timber cut.

From 1876 the Chippewa River Log Driving Company commenced to improve the Flambeau River and its tributaries for log driving purposes and made the streams possible to drive with some degree of certainty. From this date the lumber output increased annually until the annual product reached the enormous figures of over 300,000,000 feet of pine logs. This continued at the same rate until the year 1892 when the log crop commenced to decrease until in the year 1900 and what was once known as the greatest pine lumbering district in the state became a thing of the past.

By the Light of the Fire

William D. Gumaer is regarded by historians as the first permanent settler in Phillips, Wisconsin. He moved to the future Price County from Weyauwega, Wisconsin in 1874, accompanied by his wife Alice, and built a log cabin at a crossing on the Elk River, the site of the present city of Phillips. Here their first child was born and given the name Priscilla. What happened to the family in their first winter in the Pinery is recalled by Guamaer in an article he wrote for a promotional booklet on farm land opportunities in Price County published about 1915. The account is here slightly abridged.

On coming to what is now Phillips, in the Spring of 1874, I took a contract to cut out 1¼ miles of right-of-way running from Phillips north and completed the work the following winter. From this time to the Spring of 1876 I made my home at the Elk crossing. In the Spring of 1876 I moved up to Elk lake and located on its bank near where the Wisconsin Central depot now stands at Phillips. Here I went into the "hotel" business in company with Johnnie Brossard, a celebrated French cook for lumber camps. This place was known as "Hotel de-Gumaer."

During the Fall of 1875, when the engineers under Capt. Rich as chief engineer, were connecting the survey between Penoka Gap and the terminus of the road at 101-Worcester, I acted as messenger for him by reason of my having a good knowledge of the country lying between the two ends of the railroad, which had been started north from Stevens Point and south from the lake at Ashland and had reached 101 miles north from Stevens Point and south from Lake Superior to Penoka Gap. (The first through train run from Stevens Point to Ashland was on June 11th, 1877.) After the survey was completed he sent me over the line with parties who came from Minnesota with a view to take contracts to cut out right-of-way and make ties for the railroad.

On one of my trips, with a St. Paul party who took the contract to cut out the right-of-way from the Big Elk to the South Fork of Flambeau, I entered into a verbal contract with him to make 4,000 ties. He claimed he could not make a written contract until later, on account of his partners not being present. I took him at his word and when his teams and outfit came from St. Paul I helped him cut a road from W. T. Price's tote road, which had been cut that fall from Worcester to the head of Long Lake, and down to just below where the railroad bridge is now. I built my camp just across the river, near the rapids. That was the first camp built on the construction and he having the only team to haul supplies I depended upon it for my wants in the way of team work.

After I had my camp built and my wife and child settled in it, this contractor selfishly concluded that he had offered me too good a bargain, and about the best if not the only way he could see to get out of it was to shut down on furnishing me supplies. By doing so he hoped to starve me out of the camp. If he succeeded in starving me out he could put in a crew of his own and make ties for less money than he was to pay me. Well, he set out to do it.

In the first place he had about thirty men and in the bunch there was but two from Wisconsin—one a teamster driving his supply team and the other a cook. All the other men were from Minnesota. He had in the mean time built a set of camps up toward Deer Lake, on the west side of the railroad. They were afterwards occupied by W. H. Briggs for logging purposes. In order to get the supplies to his camp his team had to pass by my camp. So he gave both his teamster and cook instructions not to sell me a pound of anything. He further ordered his teamster not to bring me a pound of supplies or deliver a message for me, but, when I wanted to be moved out he could take his team and move me.

His intention was to starve myself and family out of the camp so that he could take possession of it himself. To make sure that the teamster followed out his instructions he would ride down with him in the morning as far as my camp to see that I did not give him any messages to take to Worcester, and at night he would be on hand to see that nothing was delivered to me. I was getting pretty short of supplies sure enough, but was living in hopes every day that some one would pass through the country that I knew, or, some one that was not connected with him in any way, that I might explain my position and condition to and have them take a message to my old friend, W. F. Turner, or Alex McQueen. I knew they would see I got supplies.

My wife was the only woman in the country and everyone was a stranger to us. I would not leave her alone. Well, as it happened, this contractor got his work scattered so that he was obliged to go up the line, near Sailor Creek, cruising, and was to be away two days. He gave strict instruction to his crew, cook and teamster that on his return from his trip if he found that any one had furnished me with supplies or delivered any message for me he would discharge them. But here is what happened.

The very first night that teamster drove up to my camp door, on his way to this contractor's camp with a load of supplies as usual. My wife and I were sitting by the light of the fire—no lamp, no candles. When he stopped his team and kicked my camp door open in came a quarter of beef, 100 pounds of flour, a small hog and two sacks of groceries. He never stopped for thanks but said to his horses "get up." When he got into camp that night he and the cook made up two more big sacks of groceries, comprising everything that was necessary and hid them by the roadside. In the morning when the teamster started for Worcester for a load, he put the sacks on his team and the goods were left on my shanty floor when the teamster remarked, "You will never starve while I have a chance to grub-stake you and the little woman,"

On the contractors return he continued his habit of watching the teamster until he became discouraged and gave it up, saying, "Bill Gumaer is the hardest cuss to starve out that I ever heard of. He must live on rabbits."

That same teamster and cook would never starve if I knew it. I would grub-stake them if it took the last dime I ever expected to see.

Soon after these events there was a rush for jobs and the road was lined with people that I knew. My camp proved to be in the best location of any on the line for the operations of that winter and the next spring.

Close-up picture shows size of logs in crib piers, filled with rocks, which anchored Little Falls dam at Holcombe. Picture taken in winter when river was down and frozen (below). Lady in center is Nora Lamb, one of second cooks in camp for dam crew, but photo probably taken after dam was abandoned, about 1912. Others not identified.

Trade Tokens

Most of the big lumber companies in Wisconsin operated a "company store" near their sawmill. Here employees could purchase groceries and dry goods, but instead of using their own money, employees were encouraged to use paper script or medal tokens issued by the company bookkeeper. Wages were seldom paid until spring and the average lumberjack found this a hardship but he was compensated in part by being able to get credit at the store by using the script or medal tokens sold to him and deducted from his wages.

The lumber companies encouraged employees to buy only from the company stores. In some towns, where the company owned almost all of the facilities, the employees had little choice, and a wise man knew he was flirting with dismissal if he failed to show at least some interest in the store. Most sources agree that the company charged more than a privately-owned store, but the difference was not enough to cause a consumer revolt.

Employees at the C.M. Christiansen store in Phelps, Wisconsin, called company money "shink-shank." To the west at Winegar in Vilas County, employees of the Vilas County Lumber Company said it was called "gin-seng" money, but there is strong reason to believe that "gin-seng" was a corruption of shink-shank.

The company store is not to be confused with the small store maintained in the logging camps called the "wanigan." Moreover, the wanigan carried only a few basic items, and did not usually deal in tokens or script.

Below and on opposite pages are reproductions, life-size, of script and medal coins used by several of the leading lumber companies of Wisconsin. Circumstantial evidence suggests that the money was not used before 1890 and seldom after 1920. The script and medal tokens shown here are from the collection of Gerald Johnson, leading collector of lumberjack money in Wisconsin, who lives in Wisconsin Rapids.

Left: Face of five cent note issued by Knapp, Stout & Company which was good at company store in Rice Lake. Below: Reverse side of five dollar script issued by Knapp, Stout & Company.

H. Marcus & Sons, Muscoda

C.M. Christiansen Co., Phelps

Mason-Donaldson Lumber Company
State Line (Land O' Lakes)

Vilas County Lumber Company
Winegar (or as Fosterville)

Charles W. Fish Lumber Company, Elcho

Yawkey-Bissell Lumber Company, Hazelhurst

Connor Lumber & Land Company, Laona

PICTURE CREDITS

4—Author's collection
7—Chippewa Valley Museum, Inc., Eau Claire
9—Charles Macnamara Collection, Ontario Archives, Canada. Courtesy Mrs. Jean Cunningham
10—Rusk County Historical Society
11—State Historical Society of Wisconsin (hereafter as SHSW)
12—Tony Wise, Hayward
14—A, Henry Munich, Bloomer
B, Henry Munich
15—Eau Claire Public Library
16—A, Eugene Harm, Cadott
B, Eugene Harm
17—A, Eau Claire Public Library
B, Author's collection
18—Eau Claire Public Library
19—Eau Claire Public Library
20—A, Eau Claire Public Library
B, Winona County Historical Society, Winona, MN
21—A, Joyce Gannon, Cadott
B, Eau Claire Public Library
22—A, Eau Claire Public Library
B, Eau Claire Public Library
24—Joyce Gannon
27—Eau Claire Public Library
28—Minnesota Historical Society, St. Paul, MN
29—Mabel Tainter Memorial Library, Menonmonie
30—Eau Claire Public Library
31—Eau Claire Public Library
32—Eau Claire Public Library
33—Harry Curran, Rib Lake
35—A, Randall E. Rohe, Menasha
B, Author's collection
C, Author's collection
36—A, SHSW
B, Mabel Tainter Memorial Library
37—A, Mabel Tainter Memorial Library
B, Joyce Gannon
38—Eau Claire Public Library
39—Henry Munich
40—Joyce Gannon
41—Eau Claire Public Library
42—Eau Claire Public Library
43—A, Eau Claire Public Library
B, Eau Claire Public Library
44—A, Eau Claire Public Library
B, Eau Claire Public Library
45—Chippewa Valley Museum, Inc.
46—Eau Claire Public Library
47—A, Eau Claire Public Library
B, Eau Claire Public Library
48—A, Henry Munich
B, Eau Claire Public Library
49—Arnold H. Boury, Chippewa Falls
50—Eldon Marple, Hayward
51—Runk Collection, Minnesota Historical Society
52—Timber Producers Association of Michigan and Wisconsin, Tomahawk
54—Area Research Center, University of Wisconsin-La Crosse (hereafter as ARC-La Crosse)
56—A, Eau Claire Public Library
B, Eau Claire Public Library
57—A, Eau Claire Public Library
B, Eau Claire Public Library
58—A, Eau Claire Public Library
B, Eau Claire Public Library
59—Minnesota Historical Society
60—Eau Claire Public Library
62—Mabel Tainter Memorial Library
64—Eau Claire Public Library
65—A, Chippewa Valley Museum, Inc.
B, Author's collection
66—A, SHSW
B, SHSW
68—SHSW
70—SHSW
71—SHSW
72—A, Eugene Harm
B, Minnesota Historical Society
74—A, Eau Claire Public Library
B, Minnesota Historical Society
75—A, Minnesota Historical Society
B, Mrs. Leslie Jones, Holcombe
77—Chippewa Falls Public Library
78—A, Minnesota Historical Society
B, Eau Claire Public Library
79—A, Chippewa Valley Museum, Inc.
B, Author's collection
80—Eau Claire Public Library
81—Eau Claire Public Library
82—Eau Claire Public Library
83—Eau Claire Public Library
84—A, SHSW
B, Howard Peddle, Iron River
85—SHSW
86—Author's collection
87—A, Author's collection
B, Eugene Harm
88—SHSW
89—Eau Claire Public Library
90—A, Author's collection
B, Eau Claire Public Library
91—Author's collection
92—Howard Peddle
93—William H. Gilmore, Durand
94—William H. Gilmore

95—William H. Gilmore
96—Joe Joas, Chippewa Falls
97—Randall E. Rohe
98—Smithsonian Institution, Washington, D.C.
99—A, Msgr. Stephen Anderl, Durand
B, Chippewa Valley Museum, Inc.
100—Area Research Center, University of Wisconsin-Eau Claire
103—A, Chippewa Valley Museum, Inc.
B, Chippewa Valley Museum, Inc.
105—A, *The History of Northern Wisconsin* (1881)
B, Eau Claire Public Library
106—A, Anastasia Bacher, Eau Claire
B, Anastasia Bacher
109—Charles E. Twining, Ashland
110—Chippewa Valley Museum, Inc.
111—A, Chippewa Valley Museum, Inc.
B, Chippewa Valley Museum, Inc.
112—Chippewa Valley Museum, Inc.
113—Eau Claire Public Library
115—Eau Claire Public Library
117—William H. Gilmore
118—A, Chippewa Valley Museum, Inc.
B, Chippewa Valley Museum, Inc.
C, Eau Claire Public Library
119—A, Author's collection
B, Author's collection
C, Joe Joas
120—Joe Joas
122—A, Mabel Tainter Memorial Library
B, Mabel Tainter Memorial Library
123—A, Mabel Tainter Memorial Library
B, Mabel Tainter Memorial Library
124—A, Author's collection
B, Eau Claire Public Library
125—Mrs. Philip Merrill, Taylor
126—Chippewa Valley Museum, Inc.
127—Author's collection
128—Eau Claire *Leader-Telegram*
131—Author's collection
132—*The History of Northern Wisconsin*
135—Eau Claire Public Library
136—Author's collection
137—A, Robert P. Rusch, Rib Lake
B, Robert P. Rusch
139—A, John M. Russell, Menomonie
B, Chippewa Valley Museum, Inc.
141—SHSW
142—Author's collection
145—Msgr. Stephen Anderl
146—John M. Russell
148—Maurice Staudecker, Holcombe
149—Randall E. Rohe
152—A, Henry Plagge, Holcombe
B, Henry Plagge
153—A, Author's collection
B, Eugene Harm
154—A, Mrs. Leslie Jones
B, Zac Jardine, Verona
C, Eau Claire Public Library
156—Zac Jardine
157—Henry Plagge
158—Northern States Power Company, Eau Claire
160—Zac Jardine
162—William H. Gilmore
165—A, Department of Natural Resources, Madison
B, Department of Natural Resources
166—A, Marian Carroll, Holcombe
B, Zac Jardine
167—Joyce Gannon
168—A, Zac Jardine
B, Zac Jardine
169—A, Zac Jardine
B, Zac Jardine
170—A, Marian Carroll
B, Zac Jardine
173—Department of Natural Resources
174—Joyce Gannon
175—Author's collection
176—Randall E. Rohe
179—Zac Jardine
180—Mrs. Leslie Jones
192—SHSW
193—Mrs. Leslie Jones
194—Zac Jardine
196—A, Eugene Harm
B, Eugene Harm
197—A, Eugene Harm
B, Joyce Gannon
C, Eugene Harm
198—A, Eugene Harm
B, Author's collection
200—Michael F. Slasinski, Saginaw, MI
201—Msgr. Stephen Anderl
203—Msgr. Stephen Anderl
204—Winona County Historical Society
205—Winona County Historical Society
206—Winona County Historical Society
207—William H. Gilmore
209—A, Eau Claire Public Library
B, ARC-La Crosse
211—Minnesota Historical Society
212—ARC-La Crosse
213—Winona County Historical Society
215—A, Msgr. Stephen Anderl
B, ARC-La Crosse

216—ARC-La Crosse
217—ARC-La Crosse
218—ARC-La Crosse
220—ARC-La Crosse
223—A, ARC-La Crosse
B, Winona County Historical Society
224—William H. Gilmore
225—John M. Russell
227—Author's collection
232—William H. Gilmore
234—Msgr. Stephen Anderl
235—Author's collection
239—Robert J. Scherer
241—A, William H. Gilmore
B, William H. Gilmore
244—William H. Gilmore
245—Mrs. Anna Bolz, Lugerville
246—R.C. Brown, Eau Claire
248—Author's collection
249—A, R.C. Brown
B, R.C. Brown
251—A, R.C. Brown
B, Eau Claire Public Library
252—A, Glenwood Rast, Glidden
B, Eau Claire Public Library
254—SHSW
256—Catherine Mess, Phillips
257—A, Karl & Robert Mess, Phillips
B, Mrs. Art Johnson, Phillips
258—Mrs. Art Johnson
259—Author's collection
260—A, Catherine Mess
B, Mrs. Art Johnson
261—Selm Turner, Phillips
264—A, Violet Lowe Christopherson, Phillips
B, Violet Lowe Christopherson
265—A, Violet Lowe Christopherson
B, Mrs. Anna Bolz
266—A, Catherine Mess
B, Mrs. Anna Bolz
C, Violet Lowe Christopherson
267—A, Mrs. Anna Bolz
B, Mrs. Anna Bolz
268—A, Eldon Marple
B, Marian Carroll
269—Marian Carroll
271—Minnesota Historical Society
273—Eau Claire Public Library
274—Author's collection
275—Map by David Ninneman
276—A, Author's collection
B, Henry Munich
277—Eldon Marple
278—Eldon Marple
279—Eau Claire Public Library
282—Mrs. Oscar Richter, Manitowoc
284—Author's collection
286—SHSW
289—Zac Jardine
290-91—Jerald Johnson, Wisconsin Rapids
292-93—Langlade County Historical Society

BIBLIOGRAPHY

Anderson-Sannes, Barbara. *Alma on the Mississippi 1848-1932.* Alma, WI: Alma Historical Society, 1980.

Bill, Fred A. *Navigation on the Chippewa River in Wisconsin.* Pamphlet compiled by Durand Public Library from series of articles, beginning 19 Nov., 1931, in Durand *Courier-Wedge,* a weekly newspaper.

Blair, Walter A. *A Raft Pilot's Log.* Cleveland, OH: The Arthur H. Clark Company, 1930.

Boyd, Robert T. "Up and Down the Chippewa River." *Wisconsin Magazine of History,* Vol. XIV, 1930-31, pp. 243-261.

Bunnell, Lafayette Houghton. *Winona and its Environs on the Mississippi River Ancient and Modern Days.* Winona, MN: Jones & Kroeger Company, 1897.

Corrigan, George A. *Calked Boots and Cant Hooks.* Saxon, WI: n.p., 1970.

Daniel Shaw Lumber Co. Papers in William Warren Bartlet Papers. Eau Claire Public Library.

Farquhar, F.O. *Report of the Examination of the Saint Croix and Chippewa Rivers.* U.S. 43rd Cong. 2nd sess. H. Rep. 75(6) Washington, D.C.: GPO, 1874-75.

Forrester, George, ed. *Historical and Biographical Album of Chippewa Valley.* Chicago, IL: A. Warner, 1891.

Fries, Robert F. *Empire in Pine: The Story of Lumbering in Wisconsin 1830-1900.* Madison, WI: State Historical Society of Wisconsin, 1951.

"The Mississippi River Logging Company and the Struggle for Free Navigation of Logs 1865-1900." *Mississippi Valley Historical Review.* Vol. XXXV. No. 3 December 1948, pp. 429-448.

Gates, Paul W. "Weyerhaeuser and the Chippewa Logging Industry." In *John H. Hauberg Historical Essays.* Ed. O. Fritiof Ander. Rock Island, IL: Augustana Book Concern, 1954.

Gates, Wilbur H. *The Wisconsin Pine Lands of Cornell University: A Story of Land Policy and Absentee Ownership.* Madison, WI: State Historical Society of Wisconsin, 1965.

Hedy, Ralph W., Frank E. Hill and Allen Nevins. *Timber and Men: The Weyerhaeuser Story.* New York: Macmillan, 1961.

Hurst, James Willard, *Law and Economic Growth; The Legal History of the Lumber Industry in Wisconsin.* Cambridge: Harvard University Press, 1964.

Ingram, Orrin H. *Autobiography.* Eau Claire, WI: n.p., 1912.

Kohlmeyer, Fred W. *Timber Roots: The Laird, Norton Story.* Winona, MN: Winona County Historical Society, 1972.

Merrick, George B. *Old Times on the Upper Mississippi: the Recollections of a Steamboat Pilot from 1855 to 1863.* Cleveland, OH: The Arthur H. Clark Company, 1909.

North, Matthew. *The Mississippi River Logging Company.* Winona, MN: n.p., 1912.

Peterson, Dale A. "Lumbering on the Chippewa: the Eau Claire Area 1845-1885." Diss. University of Minnesota, 1970.

Petersen, William J. *Steamboating . . . On the Upper Mississippi.* Iowa City, IA: The State Historical Society of Iowa, 1968.

Raney, William R. "Pine Lumbering in Wisconsin." *Wisconsin Magazine of History.* Vol. XIX, 1935-36, pp. 71-90.

Reynolds, A.R. *The Daniel Shaw Lumber Company: A Case Study of the Wisconsin Lumbering Frontier.* New York: New York University Press, 1957.

Russell, Charles Edward. *A-Rafting on the Mississip'.* New York: The Century Company, 1928.

Shaw, John. "Shaw's Narrative." *Wisconsin Historical Collections,* Vol. II (1856), pp. 197-232.

Sieber, George. "Sawmilling on the Mississippi: The W.J. Young Lumber Company." Diss. University of Iowa, 1969.

Twining, Charles E. "Lumbering on the Chippewa River." Diss. University of Wisconsin-Madison, 1963.

Vinette, Bruno. *"Early Lumbering on the Chippewa."* *Wisconsin Magazine of History.* Vol. IX No. 4, 1925-26, pp. 442-447.

INDEX

D

E

F

G

H

Q

R

S

T

U

V

W

Y

Z